The
Science Magpie

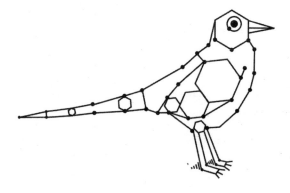

**A hoard of fascinating facts, stories,
poems, diagrams and jokes,
plucked from science
and its history**

SIMON FLYNN

ICON

Published in the UK in 2012 by
Icon Books Ltd, Omnibus Business Centre,
39–41 North Road, London N7 9DP
email: info@iconbooks.net
www.iconbooks.net

Sold in the UK, Europe and Asia
by Faber & Faber Ltd, Bloomsbury House,
74–77 Great Russell Street,
London WC1B 3DA or their agents

Distributed in the UK, Europe and Asia
by TBS Ltd, TBS Distribution Centre, Colchester Road,
Frating Green, Colchester CO7 7DW

Distributed in Australia and New Zealand
by Allen & Unwin Pty Ltd,
PO Box 8500, 83 Alexander Street,
Crows Nest, NSW 2065

Distributed in South Africa by Book Promotions,
Office B4, The District, 41 Sir Lowry Road,
Woodstock 7925

Distributed in Canada by Penguin Books Canada,
90 Eglinton Avenue East, Suite 700,
Toronto, Ontario M4P 2YE

ISBN: 978-184831-416-0

Typeset in ITC Stone by Marie Doherty

Printed and bound by
CPI Group (UK) Ltd, Croydon, CR0 4YY

For Kate and Alice

ABOUT THE AUTHOR

Simon Flynn studied chemistry at the University of Bristol and then did an MA in philosophy at York. He worked in publishing for fifteen years and is now training to be a science teacher. He lives in North London with his wife and step-daughter.

CONTENTS

INTRODUCTION

science, *n.*

5.b. Science. In modern use, often treated as synonymous with 'Natural and Physical Science', and thus restricted to those branches of study that relate to the phenomena of the material universe and their laws …

magpie, *n.* and *adj.*

B. *adj.* (*attrib.*)

1. Magpie-like: with allusion to the bird's traditional reputation for acquisitiveness, curiosity, etc.; indiscriminate, eclectic, varied.

Oxford English Dictionary

Science. What does that word conjure up in your mind? Writing over two hundred years ago, the great German authors Friedrich Schiller and Johann Wolfgang von Goethe gave this rather neat (if slightly silly-sounding) description in their collection of poetic epigrams, *Xenien*:

To one, it's a high, heavenly goddess. To another it's a cow that provides them with butter.

You might want to read that again. What they appear to be saying is that science can be viewed in terms of powerful *ends* (I'm thinking of the very best metaphorical butter here) and almost spiritual, aesthetic *means*.

In Schiller and Goethe's eyes, at least, people's responses tend to depend on which of these they focus on. But why not consider both? Why not acknowledge that one reason why science is so special is that both these aspects can be true?

There is more though. Science isn't just about laws, theories, formulae, processes and experiments. At heart it's a human activity. Without the incredible individuals who have occupied themselves with uncovering the workings of nature and applying them to our benefit where would we be? That's the last rhetorical question, I promise.

This miscellany is intended to showcase just some of the many and varied facets of science, plenty of which are unashamedly and idiosyncratically human. It's people who provide great stories and heart-warming anecdotes. Partly for this reason, the story of Erasto Mpemba, an African schoolboy who didn't give up in his quest to understand something (and who has a physical effect named after him as a result), and Darwin's note on the pros and cons of marriage are among my favourite entries in these pages.

Although this book is a miscellany, certain themes and ideas echo through its pages. One of these is mathematics. You may have noticed the '…' at the end of the definition of science given at the start of this introduction. That's because I cheated slightly – the definition actually continues 'sometimes with implied exclusion of pure mathematics'. However, this magpie sides firmly with Roger Bacon, who said 'mathematics is the door and the key to the sciences' and, as such, it is included. At least I'm being honest.

A non-scientific realm that features rather heavily in the book is the arts, and poetry in particular. In contrast with science, artistic expression of any kind is typically felt to possess a power beyond the material. I'm not sure quite how fair a view of science this is. But I believe that an appreciation of *both* realms of human endeavour can only be for the good, given the great wonders each offers. So, as writers are often told, 'show, don't tell'. That has been the intention in those entries in the book where the two cultures overlap, such as 'The poets' scientist' and Siv Cedering's 'Letter from Caroline Herschel'.

And so to the final big theme in *The Science Magpie*. The splendid short story writer Katherine Mansfield once wrote, it is of 'immense importance to learn to laugh at ourselves' and I couldn't agree more. From a smattering of groan-inducing jokes in some of the boxed fillers, to moments of more delicate wit from practitioners of science in entries such as 'Is Hell exothermic or endothermic?' and 'A chemic union', science's lighter side is very much on show because, as another writer, Colette, put it, an 'absence of humour renders life impossible'.

Science has the power to enrich people's lives, both figuratively and practically. What follows is a celebration of it, warts and all. I hope you enjoy it.

Simon Flynn, 2012

P.S.

In case it has been a while since you last did, or read, any science, I've put together a brief back-to-school appendix (see page 256), which will hopefully remind you of some of the science you were taught at school and which may be useful to recall when reading this book. Don't worry, there won't be any exam at the end of it.

'HYMN TO SCIENCE'

It seems only fitting to open *The Science Magpie* with a 'Hymn to Science', which first appeared in *The Gentleman's Magazine* in 1739 when its author, Mark Akenside, was only seventeen. The son of a butcher, it was around this time Akenside switched from preparing for a life as a nonconformist clergyman to training in medicine. He soon became a member of the Medical Society, eventually securing the position of physician to the Queen a little over twenty years later. Akenside wrote poetry throughout his life, including continuously revising his most famous work, *The Pleasures of the Imagination*, which Dr Johnson described as 'an example of the great felicity of genius'.

From the 'Hymn to Science'

Science! thou fair effusive ray
From the great source of mental Day,
*   Free, generous, and refin'd!*
Descend with all thy treasures fraught,
Illumine each bewilder'd thought,
*   And bless my lab'ring mind.*

But first with thy resistless light,
Disperse those phantoms from my sight,
*   Those mimic shades of thee:*
The scholiast's learning, sophist's cant,
The visionary bigot's rant,
*   The monk's philosophy.*

O! let thy powerful charms impart
The patient head, the candid heart,
    Devoted to thy sway;
Which no weak passions e'er mislead,
Which still with dauntless steps proceed
    Where Reason points the way.

Give me to learn such secret cause;
Let number's, figure's, motion's laws
    Reveal'd before me stand;
These to great Nature's scenes apply,
And round the globe, and thro' the sky,
    Disclose her working hand.

Next, to thy nobler search resign'd,
The busy, restless, human mind
    Thro' every maze pursue;
Detect Perception where it lies,
Catch the Ideas as they rise,
    And all their changes view.

Say from what simple springs began
The vast ambitious thoughts of man,
    Which range beyond controul,
Which seek Eternity to trace,
Dive thro' th' infinity of space,
    And strain to grasp The Whole.

[...]

> **There is no science without fancy**
> **and no art without facts.**
>   **Vladimir Nabokov (1899–1977)**

HOW DO THEY DO IT?

Astronomers do IT in the dark.

Mathematicians do IT in numbers.

Biologists do IT in the field.

Chemists do IT periodically on the table.

Geologists do IT in folded beds.

Palaeontologists do IT in the dirt.

Computer scientists do IT bit by bit.

Electrical engineers do IT until it hertz.

Physicists do IT with force.

When seismologists do IT, the Earth shakes.

Zoologists do IT with animals.

Quantum physicists do IT uncertainly.

Polymer chemists do IT in chains.

Cosmologists do IT with a bang.

Theorists do IT on paper.

Geneticists do IT in their genes.

Statisticians do IT with 99% confidence.

Planetary scientists do IT while gazing at Uranus.

Philosophers only think about doing IT.

What is IT? Why, science of course. And shame on you if you thought otherwise.

Adapted from Jupiter Scientific (www.jupiterscientific.org)

THE SCIENTIST

Have you ever wondered why we use the word 'scientist' to describe someone who works in science? Probably not – after all, it seems a pretty obvious term to use. In fact, however, its selection by the scientific community was made relatively late in the day and was a matter of some controversy.

At the 1833 meeting of the British Association for the Advancement of Science (BAAS), which had been founded just two years earlier, the English Romantic poet and polymath Samuel Taylor Coleridge raised the question of what to call someone who worked 'in the real sciences'. William Whewell, then Professor of Mineralogy at the University of Cambridge and an ordained priest, put forward the word 'scientist'. A year later he made a more public proposal when anonymously reviewing Mary Somerville's *On the Connexion of the Physical Sciences* in *The Quarterly Review*:

> *Science ... loses all traces of unity. A curious illustration of this result may be observed in the want of any name by which we can designate the students of the knowledge of the material world collectively. We are informed that this difficulty was felt very oppressively by the members of the British Association for the Advancement of Science, at their meetings at York, Oxford and Cambridge in the last three summers ...* Philosophers *was felt to be too wide and too lofty a*

term, …; savans *was rather assuming, …; some ingenious gentleman [William Whewell is modestly referring to himself here] proposed that, by analogy with artist, they might form scientist, and added that there could be no scruple in making free with this termination when we have such words as* sciolist, economist, *and* atheist – *but this was not generally palatable.*

However, Whewell persisted and in 1840 wrote in *The Philosophy of the Inductive Sciences*:

We need very much a name to describe a cultivator of science in general. I should incline to call him a Scientist.

Fierce opposition remained. The Cumbrian geologist Adam Sedgwick scribbled in the margin of his copy of Whewell's proposal 'Better die of this want [of a term] than bestialize our tongue by such a barbarism!'

Although use of the word 'scientist' dramatically increased over the following years, science practitioners didn't adopt the term immediately and even more than 50 years later someone as eminent as the biologist, and erstwhile president of the Royal Society, T. H. Huxley would write 'to any one who respects the English language, I think "Scientist" must be about as pleasing a word as "Electrocution"'.

VALUE JUDGEMENTS

When William Gladstone asked Michael Faraday what the practical worth of electricity was he is reported to have responded 'Why, sir, there is every probability that you will soon be able to tax it!' Science, as we know, is many things to many people but, as Faraday's comment implies, to politicians it's typically something to get excited about only when it's clear what its contribution to the country's coffers might be. Ditto a private company and its shareholders. However, as Homer Adkins' witty quip that 'basic research is like shooting an arrow into the air and, where it lands, painting a target' shows, it's not always obvious what areas of research within science will provide a return. Judging investment is a difficult task.

The bar chart below details the fifteen countries with the highest percentage of GDP spent on R&D in 2008 as given by the World Bank, along with a selection of countries from further down the list. Are you surprised to see the top three are Israel, Finland and Sweden? That R&D is usually important to developed countries, along with the fact it's of relatively low priority to less developed nations, is clear. But which is the chicken and which is the egg may be harder to determine.

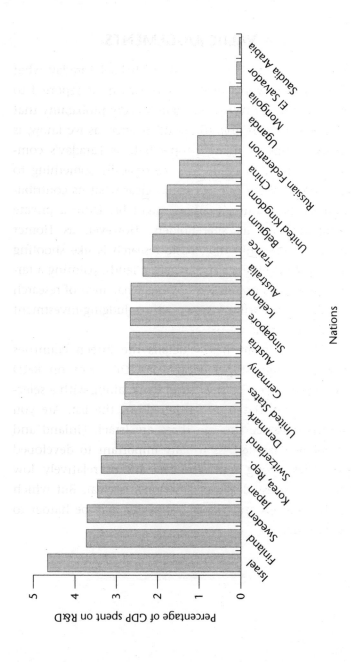

Nations

Percentage of GDP spent on R&D

5 — 4 — 3 — 2 — 1 — 0

Israel
Finland
Sweden
Japan
Korea, Rep.
Switzerland
Denmark
United States
Germany
Austria
Singapore
Iceland
Australia
France
Belgium
United Kingdom
Russian Federation
China
Uganda
Mongolia
El Salvador
Saudi Arabia

THE TRUE MEASURE OF THINGS (PART 1)

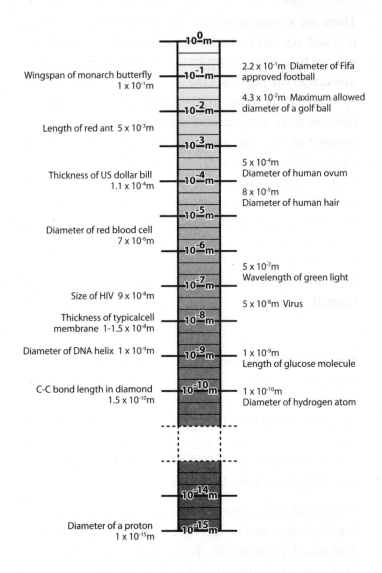

	10^0 m
Wingspan of monarch butterfly 1×10^{-1}m	10^{-1} m — 2.2 × 10^{-1}m Diameter of Fifa approved football
	10^{-2} m — 4.3 × 10^{-2}m Maximum allowed diameter of a golf ball
Length of red ant 5×10^{-3}m	10^{-3} m
Thickness of US dollar bill 1.1×10^{-4}m	10^{-4} m — 5 × 10^{-4}m Diameter of human ovum
	— 8 × 10^{-5}m Diameter of human hair
	10^{-5} m
Diameter of red blood cell 7×10^{-6}m	10^{-6} m
	— 5 × 10^{-7}m Wavelength of green light
	10^{-7} m
Size of HIV 9×10^{-8}m	— 5 × 10^{-8}m Virus
Thickness of typicalcell membrane 1-1.5×10^{-8}m	10^{-8} m
Diameter of DNA helix 1×10^{-9}m	10^{-9} m — 1 × 10^{-9}m Length of glucose molecule
C-C bond length in diamond 1.5×10^{-10}m	10^{-10} m — 1 × 10^{-10}m Diameter of hydrogen atom
	10^{-14} m
Diameter of a proton 1×10^{-15}m	10^{-15} m

AS EASY AS Al, Be, Cs

There are a great many popular songs inspired by love, loss and, oh, did I mention love, but only one can claim to have been inspired by an object of adoration quite as unusual as the periodic table of elements.

Tom Lehrer, born in 1928, is a retired mathematician, lecturer and satirical songwriter who released a number of very successful albums in the 50s and 60s. One of these included 'The Elements'. It's a song version of the then 102-element periodic table to the tune of the Major-General's Song from *The Pirates of Penzance* by Gilbert and Sullivan. Versions by various people can be found online, including one by Daniel Radcliffe, the actor who played Harry Potter. However the best undoubtedly remains the live version sung by Lehrer himself.

*

Now if I may digress momentarily from the mainstream of this evening's symposium, I'd like to sing a song, which is completely pointless but it's something I picked up during my career as a scientist. This may prove useful to some of you some day, perhaps in a somewhat bizarre set of circumstances. It's simply the names of the chemical elements set to a possibly recognizable tune.

There's antimony, arsenic, aluminum, selenium
And hydrogen and oxygen and nitrogen and rhenium
And nickel, neodymium, neptunium, germanium
And iron, americium, ruthenium, uranium

Europium, zirconium, lutetium, vanadium
And lanthanum and osmium and astatine and radium
And gold and protactinium and indium and gallium
And iodine and thorium and thulium and thallium

There's yttrium, ytterbium, actinium, rubidium
And boron, gadolinium, niobium, iridium
And strontium and silicon and silver and samarium
And bismuth, bromine, lithium, beryllium, and barium

(*Isn't that interesting? [Audience laughs] I knew you would.*
I hope you're all taking notes, because there's going to be a
short quiz next period)

There's holmium and helium and hafnium and erbium
And phosphorus and francium and fluorine and
    terbium
And manganese and mercury, molybdenum,
    magnesium
Dysprosium and scandium and cerium and caesium
And lead, praseodymium, and platinum, plutonium,
    palladium, promethium, potassium, polonium
And tantalum, technetium, titanium, tellurium
And cadmium and calcium and chromium and curium

There's sulfur, californium, and fermium, berkelium
And also mendelevium, einsteinium, nobelium
And argon, krypton, neon, radon, xenon, zinc, and
    rhodium
And chlorine, carbon, cobalt, copper, tungsten, tin, and
    sodium

These are the only ones of which the news has come to
  Ha'vard

And there may be many others but they haven't been
  discavard

THE SQUARE OF SCIENTIFIC DELIGHTS

London's Leicester Square has long been regarded as
a centre of entertainment. But what might come as a
surprise is that of the various diversions that have been
found there through history, quite a few have been scientific in nature.

On 4 February 1775, the *Morning Post and Daily
Advertiser* ran the following advert on its front page:

Museum, Leicester House, Feb. 3, 1775

*Mr. Lever's Museum of Natural and other Curiosities,
consisting of beasts, birds, fishes, corals, shells, fossils
extraneous and native, as well as many miscellaneous
articles in high preservation, will be opened on Monday
13th of February, for the inspection of the public. [...]
As Mr. Lever has in his collection some very curious
monkies and monsters, which might disgust the Ladies,
a separate room is appropriated for their exhibition, and
the examination of those only who chuse it.*

The Holophusicon ('embracing all of nature'), as it was
sometimes called, was located at Leicester House on the
northern side of Leicester Square. It held the natural history collection of Sir Ashton Lever, which visitors could

view for half a guinea. Totalling around 27,000 objects, the collection included a hippopotamus, an elephant, hummingbirds, pelicans, peacocks, bats, lizards and scorpions as well as many artifacts picked up during the explorer James Cook's second and third voyages. Writing in 1778 to her cousin Frances (author of the bestselling *Evelina*), Susan Burney described the infamous room that might 'disgust the ladies' as

> *... full of monkeys – one of which presents the company with an Italian Song – another is reading a book – another, the most horrid of all, is put in the attitude of* Venus de Medicis, *and is scarce fit to be looked at.*

The collection had initially been shown at Lever's country house, Alkrington Hall, near Manchester, before moving to London because it wasn't making enough money to sate Lever's addiction for collecting. Unfortunately, the London museum couldn't sustain itself either, despite being extremely popular and visited by George III and the Prince of Wales. Lever ended up having to sell it by lottery in 1786 (Lever offered the collection to the British Museum first but they sadly declined and it ended up fragmenting). Never quite able to cope with this loss, Lever committed suicide less than two years later.

In 1783, while the Holophusicon was still going strong, the anatomist John Hunter began renting a large house in the square, where he was able to run a medical school and museum. Hunter was another prodigious collector, and people now had the opportunity to view skeletons of kangaroos from one of Cook's voyages, as well as that of Charles Byrne, an Irishman measuring 7' 7". The acquisition of the latter, which cost Hunter the equivalent of £50,000 in today's money, is at the centre of Hilary Mantel's 1998 novel *The Giant, O'Brien*. Unlike Lever's collection, Hunter's was thankfully bought by the government after his death and now forms the core of the Hunterian Museum at the Royal College of Surgeons in London.

Seventy years later, in the aftermath of the hugely popular Great Exhibition of 1851, Leicester Square witnessed the creation of its most ambitious scientific establishment yet. Sadly, it also proved to be its last. Dominating the eastern side of the square and built in the Moorish style with a 'towering minaret', 'lofty dome' and 'abundant use of chromatic decoration', The Royal Panopticon of Science and Art's aim, according to its Royal Charter, was 'to exhibit and illustrate, in a popular form, discoveries in science and art'. It opened its doors to the general public in 1854 to considerable fanfare. No expense had been spared inside either. Placed beneath the dome was a fountain, the central jet of which shot up almost 100 feet; at the entrance of the western gallery was the largest organ in England at the time and a lift (referred to as an 'ascending carriage') that could carry eight persons at a time, transported visitors to the photography gallery. Among the displays were an aurora borealis apparatus, which enabled the creation of artificial 'northern lights', a 'crystal cistern for diving' and a gas cooker (a reviewer lamented it not being put to better

The Alhambra theatre, formerly the Royal Panopticon, dominating the eastern side of Leicester square just after the square's new garden had been opened in 1874.

use by cooking his dinner). In the basement could be found lecture theatres where demonstrations were regularly given. *The Royal Panopticon* was clearly a place that you would have to visit many times in order to appreciate all the marvels it contained. It was, the *Morning Post* enthused, 'the most magnificent temple erected for the purposes of science'.

It may well have been, but few of the public paid homage. Despite its royal charter, Queen Victoria failed to grace the altar of this 'temple' and two years later it was declared bankrupt. The building was renamed The Alhambra and became a circus, then a music hall and finally a theatre before being demolished in 1936 to make way for the Odeon cinema. The organ was sold to St Paul's Cathedral and the scientific gods fled Leicester Square never to return.

VARIATIONS ON A THEME OF OCKHAM'S RAZOR

Entia non sunt multiplicanda praeter necessitatem

No more things should be presumed to exist than are absolutely necessary

The above is the standard version of Ockham's Razor, named after the 14th century English Franciscan monk, William of Ockham. It is also known as the 'principle of parsimony' and has often been held up as a useful rule of thumb with which to judge the relative merits of two competing scientific theories that predict, or account for, the same experimental results. The razor is generally understood to mean applying a preference to the one that is simpler.

However, the point of the razor isn't primarily about simplest always being best. Instead it's a call to include only what is necessary (the two often go hand in hand but it's the latter that's the driver). The problem with this is that it isn't always clear in science what is necessary or even whether all the relevant information to enable a judgment is available. As such, the principle isn't embraced universally, which explains its heuristic, rather than absolute, value.

William of Ockham appears to have written a number of versions of the principle, and many others before and since have expressed their own versions.

Aristotle:	'Nature operates in the shortest way possible.'
	'If the consequences are the same it is always better to assume the more limited antecedent.'
William of Ockham:	'Plurality is not to be posited without necessity.'
	'No plurality should be assumed unless it can be proved by reason, or by experience, or by some infallible authority.'
	'It is futile to do with more things that which can be done with fewer.'
Johannes Kepler:	'Nature uses as little as possible of anything.'
Isaac Newton:	'We are to admit no more causes of natural things than such as are both true and sufficient to explain their appearances. Therefore, to the same natural effects we must, so far as possible, assign the same causes.'
Bertrand Russell:	'Whenever possible, substitute constructions out of known entities for inferences to unknown entities.'
Ernst Mach:	'Scientists must use the simplest means of arriving at their results and exclude everything not perceived by the senses.'
Einstein:	'Everything should be made as simple as possible, but not simpler.'

Perhaps the German-American architect Ludwig Mies van der Rohe put it most succinctly of all when he said simply, 'less is more'.

> *God made the integers; all the*
> *rest is the work of Man.*
>> German mathematician
>> Leopold Kronecker

NAME THAT NUMBER

You may realise this already but there exist many different types of numbers. As such, it's possible to classify numbers much as we do plants, animals and so on. The following is a quick primer in this classification.

Natural – These are also known as counting numbers, and are 1, 2, 3 …

Integers – These are like natural numbers, but augmented by zero and negative numbers. They are always whole e.g. … –3, –2, –1, 0, 1, 2, 3 …

Rational – Any number that is an integer *or* a ratio of integers e.g. –2, –$^{33}/_{40}$, ½, 7 …

Irrational – Any number that cannot be represented by an integer or ratio of integers e.g. $\sqrt{5}$, *e* or π. Their decimals are infinite and don't repeat e.g. $\sqrt{5}$ is 2.2360679779 …

Transcendental – These are irrational numbers that are not roots of any algebraic equation with rational coefficients. *e* is a transcendental number but wouldn't be if, say, $5e^2 + 2e + 20 = 0$ (5 and 2 in this equation are rational coefficients). It's actually very difficult to prove a number is transcendental, which partly explains why *e* and π weren't shown to be so until 1873 and 1882 respectively.

Real – Any number that is rational or irrational, so all of the above are real numbers. The symbol for the set of all real numbers is $\mathbb{R}$.

Complex – These are of the form $a + bi$, where *a* and *b* are any real number and $i^2 = -1$. *a* is said to be the real part and *b* the imaginary part of the number. Regarding imaginary numbers, examples include $\sqrt{-1} = i$ and $\sqrt{-49} = 7i$.

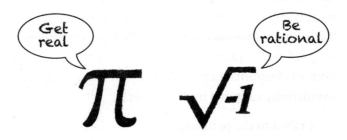

THE 'GREATEST EQUATION EVER'

Leonard Euler (pronounced 'oiler') is one of the most renowned mathematicians ever to have lived. Born in Basel, Switzerland, in 1707, his collected works fill over 70 volumes despite him losing the sight in his right eye in his early twenties and going blind in the other when 64. Euler's achievements are numerous – he solved the mathematical problem known as the Seven Bridges of Konigsberg, investigated the mathematics of music, and proved Fermat's little theorem. He was also responsible for introducing, or popularising, many bits of mathematical notation we still use today including e, π and i ($\sqrt{-1}$). In the course of his work, Euler derived the following equation:

$$e^{\pi i} + 1 = 0$$

Known as Euler's Identity, it links together what the mathematician Eli Maor has called the five most important constants in mathematics: e, i, π, 1 and 0. It has been described by the American physicist Richard Feynman as 'the most remarkable formula in mathematics', voted the most beautiful theorem in mathematics by the readers of the journal *Mathematical Intelligencer* and emerged as the joint 'greatest equation ever' (along with Maxwell's equations of electromagnetism) in a poll in *Physics World*. It is wonderfully captured in the limerick:

e raised to the pi times i,
And plus 1 leaves you nought but a sigh.
This fact amazed Euler
That genius toiler,
And still gives us pause, bye the bye.

TO MARRY OR NOT TO MARRY, THAT IS THE QUESTION

During the summer of 1838, Charles Darwin, then aged 29, found himself in a bit of a quandary. The subject worrying him was whether he should marry or not, and, if so, when. To help him come to a decision, he wrote a note listing the pros and cons of having a wife and, as would be expected from the man who would revolutionise evolutionary theory through the examination of argument and evidence, he really cut to the heart of the matter:

This is the Question

Marry

Children — (if it Please God) — Constant companion, (& friend in old age) who will feel interested in one, — object to be beloved & played with.— — better than a dog anyhow. — Home, & someone to take care of house — Charms of music & female chit-chat. — These things good for one's health. — but terrible loss of time. —

My God, it is intolerable to think of spending ones whole life, like a neuter bee, working, working, & nothing after all. — No, no won't do. — Imagine living all one's day solitarily in smoky dirty London House.— Only picture to yourself a nice soft wife on a sofa with good fire, & books & music perhaps — Compare this vision with the dingy reality of Grt. Marlbro' St.

Not Marry

*Freedom to go where one liked — choice of Society & little of it. — Conversation of clever men at clubs — Not forced to visit relatives, & to bend in every trifle. — to have the expense & anxiety of children — perhaps quarelling — **Loss of time**. — cannot read in the Evenings — fatness & idleness — Anxiety & responsibility — less money for books &c — if many children forced to gain one's bread. — (But then it is very bad for ones health to work too much)*

Perhaps my wife wont like London; then the sentence is banishment & degradation into indolent, idle fool —

Darwin's eventual conclusion?

Marry—Marry—Marry Q.E.D.,

Darwin proposed to his cousin Emma Wedgwood on 11 November 1838, writing in his journal 'the day of days!' They were married two and a half months later on 29 January 1839.

'A CHEMIC UNION'

Constance Naden (1858–1889), an admirer of the philosopher and sociologist Herbert Spencer who coined the phrase 'survival of the fittest', published two volumes of poetry during her relatively brief life. Writing in *The Speaker* in 1890, William Gladstone named her as one of the best 'poetesses' of that century in a list that also included Emily Brontë, Elizabeth Barrett Browning and Christina Rosetti.

Naden was an unusual poet for the Victorian age she lived in. For a start, she was a woman who was interested in, and who studied, science – an area assumed by many to favour the male mind. Her poetry often touched upon the realms of science and nowhere is this better demonstrated than in her witty series of four poems, 'Evolutional Erotics'. In these she presents the relationships of four couples, using the sciences to great effect in her metaphor and imagery. Below is an extract from the first in the series, which describes how a young scientist's passion for his studies is transferred to an, at best, unsuspecting, at worst, uninterested, Mary Maud Trevalyn.

From 'Scientific Wooing'

I WAS a youth of studious mind,
Fair Science was my mistress kind,
    And held me with attraction chemic;
No germs of Love attacked my heart,
Secured as by Pasteurian art
    Against that fatal epidemic.

[...]

Alas! that yearnings so sublime
Should all be blasted in their prime
    By hazel eyes and lips vermilion!
Ye gods! restore the halcyon days
While yet I walked in Wisdom's ways,
    And knew not Mary Maud Trevalyn!

[...]

I covet not her golden dower—
Yet surely Love's attractive power
    Directly as the mass must vary—
But ah! inversely as the square
Of distance! shall I ever dare
    To cross the gulf, and gain my Mary?

[...]

Bright fancy! can I fail to please
If with similitudes like these
    I lure the maid to sweet communion?
My suit, with Optics well begun,

By Magnetism shall be won,
    And closed at last in Chemic union!

At this I'll aim, for this I'll toil,
And this I'll reach—I will, by Boyle,
    By Avogadro, and by Davy!
When every science lends a trope
To feed my love, to fire my hope,
    Her maiden pride must cry "Peccavi!"

I'll sing a deep Darwinian lay
Of little birds with plumage gay,
    Who solved by courtship Life's enigma;
I'll teach her how the wild-flowers love,
And why the trembling stamens move,
    And how the anthers kiss the stigma.

Or Mathematically true
With rigorous Logic will I woo,
    And not a word I'll say at random;
Till urged by Syllogistic stress,
She falter forth a tearful "Yes,"
    A sweet "Quod erat demonstrandum*!"*

> *A modern poet has characterised the personality of art and the impersonality of science as follows: Art is I; Science is We.*
>
> **Claude Bernard (1813–1878), French physiologist**

BINARY BASICS

The number system we use in everyday life is base ten (Remember all those tens and units sums you had to do as a child?). There are many theories as to why this is the case, the most popular being that we have ten digits on our hands. Various cultures have used other counting systems, however. The Babylonians used base 60, giving us 60 minutes in an hour, but base ten dominates. Binary, however, uses base two. This means that only two symbols, typically 0 and 1, are needed to express any value.

The table shows how some familiar base ten numbers look in binary.

	2^6 or 64	2^5 or 32	2^4 or 16	2^3 or 8	2^2 or 4	2^1 or 2	2^0 or 1
The base ten numbers							
1							1
2						1	0
3 (=2+1)						1	1
4					1	0	0
5 (4+1)					1	0	1
10 (=8+2)				1	0	1	0
100 (=64+32+4)	1	1	0	0	1	0	0

The table below shows a variety of other base ten numbers up to a maximum of 127 using binary numeration.

	2^6 or 64	2^5 or 32	2^4 or 16	2^3 or 8	2^2 or 4	2^1 or 2	2^0 or 1
13 (=8+4+1)				1	1	0	1
29 (=16+2)			1	1	1	0	1
53 (=32+16+16+4+1)		1	1	0	1	0	1
127 (=64+32+16+8+4+2+1)	1	1	1	1	1	1	1

By employing the binary system, it's possible to count up to 1023 using just your two hands if a finger or thumb up stands for 1 and a finger or thumb down represents 0 (So, all ten digits up would represent 1111111111 = 1023 in base ten).

Binary is particularly useful for computers because the digits 1 and 0 can be expressed in a variety of ways such as on/off, yes/no, electric potential/no electric potential. In Douglas Adams' *A Hitchhikers Guide to the Galaxy*, the supercomputer Deep Thought gave the answer to the meaning of life as 42. However, it would probably have found it more natural to express the answer as 101010, which has a rather nice symmetry to it, don't you think? Binary has also given rise to one of the very best maths jokes:

> There are only 10 types of people in the world: those who understand binary, and those who don't.

Okay, I'll get my coat.

THE ENERGY OF RICHTER

Earthquake magnitude is typically a measurement of ground motion, which is then expressed as a value on the Richter scale, developed in 1935 by Charles Richter. It is based on powers of ten, meaning that an earthquake measuring 5 on the Richter scale has a recorded seismograph amplitude ten times greater than one measuring 4.

Magnitude can also be translated into the seismic energy released by an earthquake, measured in joules (J). Here an increase of one on the Richter scale represents an over thirty-fold increase in the amount of seismic energy.

This enables the power of earthquakes to be compared with other energy sources and vice versa.

Richter	Energy	Event
6	6.3×10^{13} J	Power of the atomic bomb dropped on Hiroshima.
6.1	9×10^{13} J	Amount of energy in 1 gram of matter according to Einstein's celebrated $E = mc^2$.
6.3	1.8×10^{14} J	Christchurch, New Zealand, earthquake, 2011.
7	2×10^{15} J	Haiti earthquake, 2010.
7.5	1×10^{16} J	Impact energy that formed meteor crater in Arizona.
8	6.3×10^{16} J	Approximate value of the 1556 Shaanxi earthquake, China – the most devastating earthquake in recorded history, estimated to have killed more than 830,000 people.

(continued)

Richter	Energy	Event
8.3	1.5×10^{17} J	Estimated energy released by the eruption of Krakatoa in 1883.
8.3	1.7×10^{17} J	Total energy from the Sun that hits the Earth every second.
8.4	2.4×10^{17} J	Tsar Bomba, the largest nuclear bomb ever detonated.
9	2×10^{18} J	Tohoku earthquake off the east coast of Japan, 2011.
9.5	1.3×10^{19} J	Total consumption of electrical energy in the US in 2009.

As you can see, the most devastating earthquake ever recorded is not the earthquake with the greatest magnitude. There are plenty of other factors which have a bearing on how much damage an earthquake causes, such as the depth of the quake (shallow ones are typically more devastating), the physical environment, population density and the quality of building construction.

> *Did we know the mechanical affections of the particles of rhubarb, hemlock, opium, and a man, as a watchmaker does those of a watch … we should be able to tell beforehand that rhubarb will purge, hemlock kill, and opium make a man sleep.*
>
> John Locke, *An Essay Concerning Human Understanding*

SCIENCE'S MOLECULE OF THE YEAR/ BREAKTHROUGH OF THE YEAR AWARD

In 1989, echoing *Time* magazine's Man of the Year, the journal *Science* began awarding the title of Molecule of the Year, 'honouring the scientific development of the year most likely to have a major impact on scientific advances and societal benefits'. If we're being pedantic, the winner wasn't always a molecule but was sometimes a process. In 1995, there was a further shift when the award was given to a 'state of matter' that had been hypothesised 70 years earlier and finally produced that year. Perhaps not surprisingly, the award was subsequently given the more all-embracing title Breakthrough of the Year. Here's a list of the winners:

Molecule of the Year

1989   PCR (polymerase chain reaction), a technique that enables many copies of DNA to be generated from a tiny amount

1990   Manufacture of synthetic diamonds

1991 C_{60} (buckminsterfullerene), an allotrope of carbon that looks a bit like a football

1992   NO (nitric oxide) and its role in how cells in the body operate

1993   p53, a protein found in the body which was recognised for its potential as a tumour suppressor

1994   The DNA repair enzyme system

1995   Bose-Einstein condensate (a state of matter)

Breakthrough of the Year

1996   Advances in the understanding of HIV disease

1997   The cloning of Dolly the sheep

1998   The ever-increasing expansion of the universe and the increasing evidence of dark matter

1999   The potential of stem cells

2000   The mapping of the human genome

2001   Nano, or molecular, circuits

2002   Small RNAs and their role in genome regulation

2003   The realisation that the universe is mostly made of dark energy

2004   The rover missions to Mars

2005   Evolution in action

2006   Proving the Poincare Conjecture

2007   Human genetic diversity i.e. looking at genomes individually

2008   Reprogramming cells through the insertion of genetic material

2009   *Ardipithecus ramidus*, a 4.4 million year-old skeleton discovered fifteen years earlier, is shown to be a human ancestor

2010   The first machine to function according to the rules of quantum mechanics is unveiled

2011   The finding that antiretroviral drugs reduce the risk of heterosexual transmission of HIV

GEOLOGICAL TIME PIECE

Sometimes, discovering when something happened only begins to make sense when it's seen in relation to other events. This is particularly easy to do if the total time is shown in the form of a clock. The age of Earth is about 4.5 billion years – on the clock, 1 second equals approximately 52,000 years and one hour is 187.5 million years.

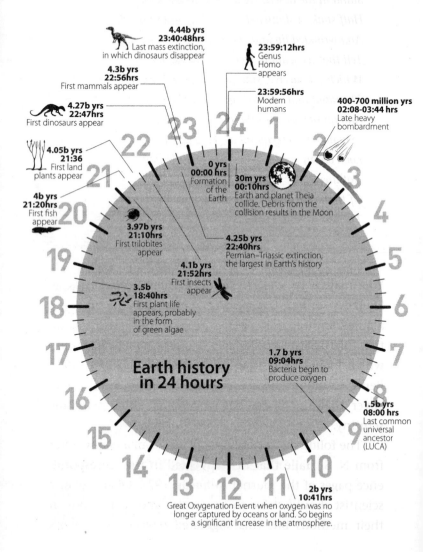

4.44b yrs
23:40:48hrs
Last mass extinction, in which dinosaurs disappear

23:59:12hrs
Genus
Homo
appears

4.3b yrs
22:56hrs
First mammals appear

23:59:56hrs
Modern
humans

4.27b yrs
22:47hrs
First dinosaurs appear

400-700 million yrs
02:08-03:44 hrs
Late heavy
bombardment

4.05b yrs
21:36
First land
plants appear

0 yrs
00:00 hrs
Formation
of the
Earth

30m yrs
00:10hrs
Earth and planet Theia
collide. Debris from the
collision results in the Moon

4b yrs
21:20hrs
First fish
appear

3.97b yrs
21:10hrs
First trilobites
appear

4.25b yrs
22:40hrs
Permian–Triassic extinction,
the largest in Earth's history

4.1b yrs
21:52hrs
First insects
appear

3.5b
18:40hrs
First plant life
appears, probably
in the form
of green algae

1.7 b yrs
09:04hrs
Bacteria begin to
produce oxygen

**Earth history
in 24 hours**

1.5b yrs
08:00 hrs
Last common
universal
ancestor
(LUCA)

2b yrs
10:41hrs
Great Oxygenation Event when oxygen was no
longer captured by oceans or land. So begins
a significant increase in the atmosphere.

'TWIN LIMB-LIKE BASALT COLUMNS'

I met a traveller from an antique land
Who said: 'Two vast and trunkless legs of stone
Stand in the desert. Near them on the sand,
Half sunk, a shattered visage lies, whose frown
And wrinkled lip and sneer of cold command
Tell that its sculptor well those passions read
Which yet survive, stamped on these lifeless things,
The hand that mocked them and the heart that fed.
And on the pedestal these words appear:
"My name is Ozymandias, King of Kings:
Look on my works, ye mighty, and despair!"
Nothing beside remains. Round the decay
Of that colossal wreck, boundless and bare,
The lone and level sands stretch far away'.

'Ozymandias', by Percy Bysshe Shelley

You may very well have encountered Percy Bysshe Shelley's sonnet 'Ozymandias' somewhere before – it's a much anthologised poem, not to mention the inspiration for The Sisters of Mercy's cracking song 'Dominion', which includes the final line of the poem in its lyrics. Being so ubiquitous it's perhaps not surprising to discover that the poem has also come under scientific scrutiny.

The following spoof is an extract from a comic letter from N. S. Haile that first appeared in the correspondence pages of the journal *Nature* in 1977. Of course any scientist should strive to be precise and meticulous in their methods but this suggested rewrite of Shelley's poem to make it acceptable for publication in a scientific journal does, I hope, show that scientists are also able to laugh at their own peculiarities:

Suggested re-written manuscript (summary)

Twin limb-like basalt columns ('trunkless legs') near Wadi Al-Fazar, and their relationships to plate tectonics

Ibn Batuta[1] and P. B. Shelley[2]

In a recent field trip to north Hadhramaut, the first author observed two stone leg-like columns 14.7m high by 1.8m in diameter (medium vast, ASTM grade scale for trunkless legs) rising from sandy desert 12.5km southwest of Wadi Al-Fazar (Grid 474 753). The rock is a tholeiitic basalt (table 1); 45 analyses by neutron activation technique show that it is much the same as any other tholeiitic basalt (table 2). A large boulder 6m southeast of the columns has been identified as of the 'shattered visage' type according to the classification of Pettijohn (1948, page 72). Granulometric analysis of the surrounding sand shows it to be a multimodal leptokurtic slightly positively skewed fine sand with a slight but persistent smell of camel dung. Four hundred and seventy two scanning electron photomicrographs were taken of sand grains and 40 are reproduced here; it is obvious from a glance that the grains have been derived from pre-cambrian anorthosite and have undergone four major glaciations, two subductions, and a prolonged dry spell. One grain shows unique lozenge-shaped impact pits and heart-like etching patterns which prove that it spent some time in upstate New York.

There is no particular reason to suppose that the columns do not mark the site of a former hotspot, mantle plume, triple junction, transform fault, or abduction zone (or perhaps all of these).

Keywords: plate tectonics, subduction, obduction, hotspots, mantle plume, triple junction, transform fault, trunkless leg, shattered visage.

[1]*School of Earth & Planetary Sciences, University of the Fertile Crescent.*
[2]*formerly of University College, Oxford.*

The scientific man is merely the minister of poetry. He is cutting down the Western Woods of Time; presently the poetry will come there and make a city and gardens. This is always so. The man of affairs works for the behoof and use of poetry. Scientific facts have never reached their proper function until they merge into new poetic relations established between man and man, between man and God, or between man and Nature ...

Sidney Lanier, 'The Legend of St. Leonor', *Music and Poetry*, 91

MNEMONIC FOR REMEMBERING
THE GEOLOGICAL TIMESCALE

For those of you who have always had a burning desire to be able to recite the periods of the geological time-scale – help is at hand in the form of this indispensable mnemonic:

Camels Often Sit Down Carefully. Perhaps Their Joints Creak. Possibly Early Oiling Might Prevent Premature Harm

Cambrian, Ordovician, Silurian, Devonian, Carboniferous, Permian, Triassic, Jurassic, Cretaceous, Palaeocene, Eocene, Oligocene, Miocene, Pliocene, Pleistocene, Holocene.

Now you need never be without a party trick!

'MOSQUITO DAY'

A year after their inception in 1901, Sir Ronald Ross (1857–1932) became the first Briton to be awarded a Nobel prize. Five years earlier, while combining his research with medical service in India, Ross had shown the *Anopheles* mosquito to be a necessary factor in the transmission of malaria. It was a finding that would have an enormous impact on the fight against the disease. For a man who had failed the licentiate examination of the Society of Apothecaries as a medical student, the Nobel prize in Physiology or Medicine 1902 and subsequent knighthood was quite a change of fortune.

A published poet and novelist – his memoirs won the James Tait Black memorial prize in 1923 – Ross wrote the following poem, now inscribed on a monument to Ross at the Presidency General Hospital, Calcutta, describing his famous discovery on 20 August 1897, which he later called 'Mosquito Day' and celebrated annually thereafter.

This day relenting God
  Hath placed within my hand
A wondrous thing; and God
  Be praised; At his command,

Seeking His secret deeds
  With tears and toiling breath,
I find thy cunning seeds,
  O million-murdering Death.

I know this little thing
  A myriad men will save.
O Death, where is thy sting?
  Thy victory, O Grave?

> **It's okay to sleep with a hypothesis, but you should never become married to one.**
>
> **Anonymous**

THE ETYMOLOGY AND SCRABBLE SCORE OF SOME COMMON SCIENTIFIC WORDS

Word	From...	Meaning	Scrabble score
Acid	Latin, *acidus*	Sour	7
Alkali	Arabic, *al qalíy*	'The roasted ashes', the name originally given to soda-ash	10
Anode	Greek, *ana* and *hodos*	Up path – positive electrode towards which current flows	6
Anthrax	Greek, *anthrax*	Charcoal – the disease causes black spots to appear on the skin	17
Argon	Greek, *argos*	Lazy, idle – argon was difficult to isolate because it's so chemically unreactive	6
Baryon	Greek, *barys*	Heavy – At the time of their discovery, baryons had the greatest mass of the subatomic particles	11
Cathode	Greek, *kata* and *hodos*	Down path – negative electrode away from which current flows	13
Cell	Latin, *cella*	Small room	6

(continued)

Word	From...	Meaning	Scrabble score
Chlorine	Greek, *chloros*	Pale green	13
Eclipse	Greek, *ek* and *leipo*	Failure to appear or leave out	11
Electron	Greek, *elektron*	Amber – when rubbed, amber becomes charged with static electricity	10
Energy	Greek, *en* and *ergon*	To work in	10
Gas	Greek, *khaos*	Chaos	4
Genetic	Greek, *genesis*	Coming into being	10
Gravity	Latin, *gravis*	Heavy	14
Hadron	Greek *adros*	Thick, bulky	10
Helium	Greek, *helios*	The Sun – helium was first discovered through spectral analysis of a solar eclipse	11
Hydrogen	Greek, *hydro* and *gene*	Water producer	16
Malaria	Italian, *mala* and *aria*	Bad air	9
Mitochondria	Greek, *mitos* and *chondros*	Thread-like grains	20
Molecule	Latin, *moles* and *culus*	Small mass	11

(continued)

Word	From...	Meaning	Scrabble score
Nucleus	Latin, *nucleus*	Kernel	9
Plankton	Greek, *planktos*	Wanderer	14
Proton	Greek, *protos*	The first	8
Quantum	Latin, *quantum*	How much	18
Science	Latin, *scientia*	Knowledge	11
Species	Latin, *species*	Appearance, kind	11

TAKING ACID

pH (short for 'power of hydrogen') is a logarithmic scale regarding the concentration of hydrogen ions (H^+) in a solution – the lower the pH, the greater the concentration of hydrogen ions. The scale is used to measure how acidic or alkaline a solution is. Pure water has a pH of 7.0, which refers to a hydrogen ion concentration of 10^{-7} mol dm^{-3} (you don't need to worry exactly what that means), and is considered neutral. Any solution with a pH less than 7 is said to be acidic and anything above, alkaline.

Because the scale is logarithmic, when the concentration of hydrogen ions changes by a factor of ten, the pH changes by only one unit. To put that in perspective we can look at the powerhouses of Britain's 18th- and 19th-century industrial revolution, such as Manchester and Huddersfield. The cities almost completely surround the peat moorland at Blakelow in the Peak District, now an English national park. A century and a half of sulfur-ous emissions from these cities have resulted in the park's peat now having a pH of 2 (a hydrogen ion concentration of 10^{-2} mol dm^{-3}), as opposed to the typical peat value of 4 (or a hydrogen ion concentration of 10^{-4} mol dm^{-3}). The difference between 2 and 4 on the scale per-haps doesn't sound like much, but in fact it means the concentration of hydrogen ions in Blakelow peat is 100 times greater than it would ordinarily be expected to be.

The pH scale was originally developed by the Danish scientist Søren Sørenson (1868–1939) while working at the Carlsberg Laboratory in Denmark. Shown overleaf are the approximate pH values of some everyday solu-tions, several of which might surprise you, especially if you consider your teeth's enamel is affected at pH 5.5 and below.

A GOOD INDICATOR OF ...

Using just a red cabbage, a kettle, a pan and some glass jars or bottles, it's possible to make a pH indicator solu-tion that will let you know whether something is acidic or alkaline, as well as indicating roughly to what degree. Here's how:

Slice and dice the red cabbage and put it in the pan. Now cover it with boiling water and give it a good stir. Let it stand for about fifteen minutes.

Use a sieve to remove the cabbage, leaving you with just the liquid. This is your indicator. It should be stored in a clean bottle, preferably in the dark.

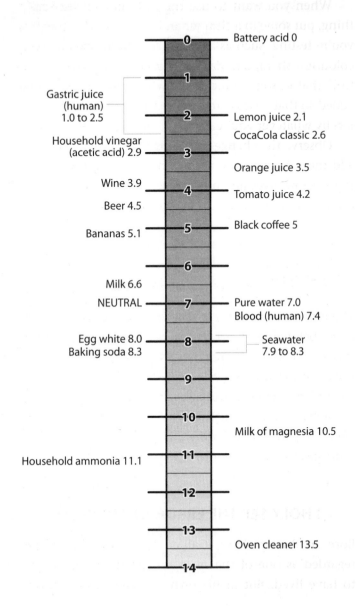

Gastric juice (human) 1.0 to 2.5

Household vinegar (acetic acid) 2.9

Wine 3.9

Beer 4.5

Bananas 5.1

Milk 6.6

NEUTRAL

Egg white 8.0
Baking soda 8.3

Household ammonia 11.1

0 — Battery acid 0

1

2 — Lemon juice 2.1
CocaCola classic 2.6

3

Orange juice 3.5

4 — Tomato juice 4.2

5 — Black coffee 5

6

7 — Pure water 7.0
Blood (human) 7.4

8 — Seawater 7.9 to 8.3

9

10 — Milk of magnesia 10.5

11

12

13 — Oven cleaner 13.5

14

When you want to use the indicator to test something, put some in a clear jar and then add whatever it is you're testing, such as apple juice, baking soda or even colourless shampoo. Make sure you always have a 'control', that is, some indicator to which nothing has been added so that it's easy to see the change caused in other jars by whatever you're testing.

Observe the change in colour that occurs when you add the substance you are testing to the indicator. The table below gives an approximate guide to the pH that the various possible colours correspond to:

Colour	Pink	Dark red/ purple	Violet	Blue	Blue-green	Green-yellow
Approximate pH	1–2	3–4	5–7	8	9–10	11–12

The indicator works because red cabbage contains a pigment belonging to a class of molecules called anthocyanins. This pigment changes colour depending on the concentration of H^+ ions, because they very slightly change its chemical structure. As such, the same species of cabbage can be planted in different locations and, depending on the pH of the different soils, the resulting cabbage crops can differ in colour.

I HOLY SEE THE ERROR OF MY WAYS

Born in Pisa in 1564, Galileo Galilei is now widely regarded as one of the most important scientists ever to have lived. But in his own time he was effectively persecuted by the Catholic Church for daring to propound what is now a generally accepted truth about the structure of the Solar System.

Galileo left university in 1585 without the degree in medicine he'd been studying for. However, since he showed an exceptional aptitude for mathematics his family were able to help him secure a professorship in this subject four years later.

By 1610 he was living in Venice, and it was in this year that *Sidereal Messenger* was published – his work describing the discoveries he had made using his own improved version of the recently-invented telescope. It quickly brought him to the attention of Europe's intelligentsia. Included in the book were the moons of Jupiter, the uneven surface of the Earth's moon and the observation that there were many more stars in the heavens than could be perceived with the naked eye.

The Church fêted Galileo for his findings the following year. However, it was becoming clear that the Ptolemaic view of the Solar System – in which the Sun, the Moon and all the planets orbited the Earth – was no longer supported by the evidence. This left two alternatives: one which had been proposed by Nicholas Copernicus almost 70 years previously and one put forward by the Danish astronomer Tycho Brahe.

If we look at Brahe's theory first. His astronomical observations led him to stick with a stationary Earth at the centre of things, but with only the Sun and the Moon orbiting it, while the planets now orbited the Sun. As far as the Church was concerned, this was okay – it was still consistent with a literal reading of the Bible (sections such as Ecclesiastes 1:5 'The Sun also arises, and the Sun goes down, and hastens to the place where it rose'), while also accounting for many of the anomalies in Ptolemy's arrangement, which had dominated thought for the previous 1,400 years.

Copernicus' system, however, had the Sun at the centre and the Earth orbiting it, along with all the other planets.

Galileo and Kepler (more on the latter later in the book) argued for the Copernican system, pretty much everyone else for Brahe's. One of the key issues in the debate was to what extent the Bible was the whole truth and nothing but the truth. Since his theory seemed to directly contradict the Bible, in 1616, the Church condemned Copernicus – any book propounding his work was banned. Galileo was effectively given a warning. However, he took the view that the Church needed to be convinced of its mistake.

Five years later, Urban VIII – formerly Maffeo Barberini – became pope. Galileo's previous book, the *Assayer*, has been dedicated to Barberini. This gave Galileo the confidence to begin a new one, *Dialogue Concerning the Two Chief World Systems: the Ptolemaic and the Copernican*, his intention being to persuade non-specialists to his view of a heliocentric (sun-centred) system. It was published in 1632. The Church's censors, the Congregation of the Index, allowed its publication, but when the pope came to read it he is reported to have 'exploded into great anger'. On 12 April 1633, Galileo was forced to stand trial. Incredibly, his defence was a denial of holding a

Copernican view. On the 22 June 1633, Galileo signed the recantation reproduced below and was sentenced to house arrest. In failing health, and going blind, he went on to write what many consider his masterpiece – *Discourses on Two New Sciences*. He died on 8 January 1642.

*

I, Galileo Galilei, son of the late Vincenzio Galilei of Florence, aged seventy years, being brought personally to judgment, and kneeling before you, Most Eminent and Most Reverend Lords Cardinals, General Inquisitors of the Universal Christian Republic against heretical depravity, having before my eyes the Holy Gospels which I touch with my own hands, swear that I have always believed, and, with the help of God, will in future believe, every article which the Holy Catholic and Apostolic Church of Rome holds, teaches, and preaches. But because I have been enjoined, by this Holy Office, altogether to abandon the false opinion which maintains that the Sun is the centre and immovable, and forbidden to hold, defend, or teach, the said false doctrine in any manner [...] I am willing to remove from the minds of your Eminences, and of every Catholic Christian, this vehement suspicion rightly entertained towards me, therefore, with a sincere heart and unfeigned faith, I abjure, curse, and detest the said errors and heresies, and generally every other error and sect contrary to the said Holy Church; and I swear that I will never more in future say, or assert anything, verbally or in writing, which may give rise to a similar suspicion of me; but that if I shall know any heretic, or any one suspected of heresy, I will denounce him to this Holy Office, or to the Inquisitor and Ordinary of the place in which I may be. I swear, moreover, and promise that I will fulfil and observe fully all the penances which have been or shall be laid on me by this Holy Office. But if it shall happen that I violate any of my said promises, oaths, and protestations (which God avert!), I subject myself to all the pains and punishments which have been decreed and promulgated by the sacred canons and other general and particular constitutions against delinquents of this description. So, may God help me, and His Holy Gospels, which I touch with my own hands, I, the above named Galileo Galilei, have abjured, sworn, promised, and bound myself as above; and, in witness thereof, with my own hand have subscribed this present writing of my abjuration, which I have recited word for word.

*

The English poet and polemicist John Milton, author of *Paradise Lost*, visited Galileo during his house arrest. This event would later feature in his *Areopagitica*, a stirring defence of free speech and attack on censorship. It's fair to say the church's actions towards Galileo would go down as one of the worst PR disasters in its history and a building block in the belief of some that there is a natural antagonism between science and religion.

RABBIT, RABBIT, RABBIT-RABBIT, RABBIT-RABBIT-RABBIT, RABBIT-RABBIT-RABBIT-RABBIT-RABBIT

In 1202, the mathematician Leonardo of Pisa, better known today as Fibonacci, published *Liber Abaci* (*The*

Book of Calculation). It was via this book that Hindu numerals (the numeric symbols we still use today) were first introduced to Europe. It also contained a problem regarding the breeding of rabbits that was to have a considerable impact on the history of mathematics. Fibonacci asked how many pairs of rabbits could be produced in a year from a single pair in an enclosed space if: each pair produced a new breeding pair every month, including the initial pair in the first month; each new pair started reproducing at the age of one month; and no rabbit ever died.

The answer, it turned out, was:

Month	1	2	3	4	5	6	7	8	9	10	11	12
Adult pairs	1	2	3	5	8	13	21	34	55	89	144	233
New pairs	1	1	2	3	5	8	13	21	34	55	89	144
Total pairs	2	3	5	8	13	21	34	55	89	144	233	377

What's interesting about the above isn't the actual answer, 377, but rather the sequence of numbers in the three rows – they're basically the same and they're what we now refer to as the Fibonacci sequence:

1, 1, 2, 3, 5, 8, 13, 21, 34, 55, 89, 144, 233, 377…

And something incredibly fascinating lies within this sequence.

If you look at the ratio of successive pairs of Fibonacci numbers placing the larger number first, (e.g. 144/89 and 233/144), as we get further along the sequence the ratio between the two gets closer and closer to 1.618 … or, more precisely, $\frac{1}{2}(1+\sqrt{5})$. This is commonly known as the golden ratio, ϕ.

The golden ratio is often cited as an aesthetically pleasing proportion, and has been used extensively in the arts during the last two and a half thousand years. Many people have used it deliberately – the French architect Le Corbusier used the ratio in some of his buildings, for example, as did Salvador Dali in his painting *The Sacrament of the Last Supper*, and for a period of time book page proportions were based on it (but not this one sadly). There are many other occasions where its appearance is more likely by chance and indeed is only approximate.

Much more startling than their use by humans, however, is the number of times the Fibonacci sequence and golden ratio appear in nature. The number of petals on many flowers is a Fibonacci number. If you look at the head of a sunflower, the florets of a cauliflower, the bumps on a pineapple, or the scales of a pinecone, you will notice a particular spiral pattern – so many in one direction and a different number in the other. Nearly always, these two numbers will be consecutive Fibonacci numbers, which we earlier saw are the best whole number approximations of ϕ. And when it comes to the leaves of a plant, ϕ crops up in the typical angle between adjacent leaves – 137.5°. Known as the 'golden angle', its connection with ϕ is: $360°/\phi \approx 222.5°$, which is the same as $360° - 137.5°$. This arrangement, which in three dimensions is spiral, is believed to give the best possible exposure to light for every leaf, taking into account their position in relation to each other. Wow!

BAN DIHYDROGEN MONOXIDE!

The following scientific spoof by Craig Jackson is his version of an idea first originated by three students at the University of California, Santa Cruz, in 1989. It's a superb example of how scientific language and terminology, along with judicious selection and disingenuous presentation of 'facts', can be used to confuse the unprepared. Everything below is, strictly speaking, true. But once you realise exactly what molecule is being discussed, any fear simply evaporates!

*

The Invisible Killer

Dihydrogen monoxide is colorless, odorless, tasteless, and kills uncounted thousands of people every year. Most of these deaths are caused by accidental inhalation of DHMO, but the dangers of dihydrogen monoxide do not end there. Prolonged exposure to its solid form causes severe tissue damage. Symptoms of DHMO ingestion can include excessive sweating and urination, and possibly a bloated feeling, nausea, vomiting and body electrolyte imbalance. For those who have become dependent, DHMO withdrawal means certain death.

Dihydrogen Monoxide:

- is also known as hydroxyl acid, and is the major component of acid rain
- contributes to the 'greenhouse effect'
- may cause severe burns
- contributes to the erosion of our natural landscape

- accelerates corrosion and rusting of many metals
- may cause electrical failures and decreased effectiveness of automobile brakes
- has been found in excised tumors of terminal cancer patients.

Contamination is Reaching Epidemic Proportions!
Quantities of dihydrogen monoxide have been found in almost every stream, lake, and reservoir in America today. But the pollution is global, and the contaminant has even been found in Antarctic ice. DHMO has caused millions of dollars of property damage in the Midwest, and recently California.

Despite the Danger, Dihydrogen Monoxide is Often Used:
- as an industrial solvent and coolant
- in nuclear power plants
- in the production of styrofoam
- as a fire retardant
- in many forms of cruel animal research
- in the distribution of pesticides – even after washing, produce remains contaminated by this chemical
- as an additive in certain 'junk-foods' and other food products.

Companies dump waste DHMO into rivers and the ocean, and nothing can be done to stop them because this practice is *still legal*. The impact on wildlife is *extreme*, and we cannot afford to ignore it any longer!

The Horror Must Be Stopped!

The American government has refused to ban the production, distribution, or use of this damaging chemical due to its 'importance to the economic health of this nation'. In fact, the navy and other military organizations are conducting experiments with DHMO, and designing multi-billion dollar devices to control and utilize it during warfare situations. Hundreds of military research facilities receive tons of it through a highly sophisticated underground distribution network. Many store large quantities for later use.

It's Not Too Late!

Act NOW to prevent further contamination. Find out more about this dangerous chemical. What you don't know can hurt you and others throughout the world.

Dihydrogen monoxide = water (H_2O)

SOLAR ECLIPSES GALORE

Throughout history, solar eclipses have captured the human imagination more than any other astronomical event. From Homer's *Odyssey* to H. Rider Haggard's *King Solomon's Mines*, they have often featured in literature as dire portents. That particular power has perhaps now passed but witnessing an eclipse is still an incredible experience. The following gives an idea of where the next few eclipses will appear.

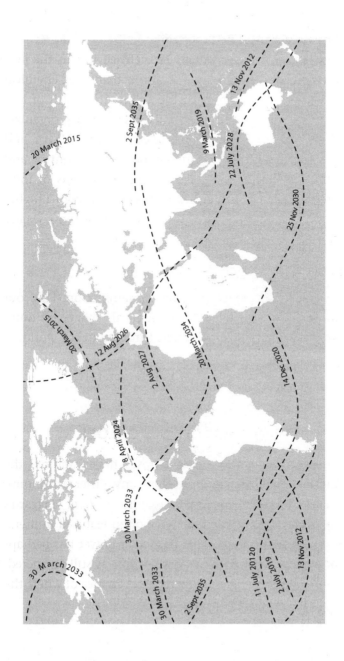

'A REMARKABLE BOOK, SURE TO MAKE A MIGHTY STIR'

Charles Darwin's *On the Origin of Species* was published on 24 November 1859. Despite costing almost as much as a policeman's weekly wage, the book sold out of its first print run of 1,250 copies in a single day.

A tome of almost mythic status in the history of science, its initial critical reception makes interesting reading. Out of the daily newspapers, only *The Times*, *Morning Post* and the *Daily News* reviewed it, along with a similar number of weeklies. The *News of the World* contained a notice with an extract from the book on slave-making ants! But the review journals, which played a vital role in public debate, covered the book in significant force. What follows are the opening two paragraphs of a review that appeared in the *Examiner* on 3 December 1859, less than two weeks after the book was published. One can't help but admire the reviewer's even-handedness, particularly as three weeks later Thomas Henry Huxley, later known as Darwin's Bulldog because of his vociferous support of Darwin's theory, wrote in his review in *The Times* that it was currently impossible to 'affirm absolutely either the truth or falsehood of Mr. Darwin's views' and that it could take another twenty years to determine whether Darwin was right or had, in fact, 'over-estimate[d] the value of his principle of natural selection' (privately though, Huxley is said to have remarked 'how stupid of me not to have thought of that' after he first read Darwin's book).

*

'On the Origin of Species by Means of Natural Selection, or the Preservation of Favoured Races in the Struggle for Life. By Charles Darwin, M.A., F.R.S., Author of 'Journal of Researches During a Voyage Round the World.' John Murray, Albermarle Street, 1859.

'This is a remarkable book, sure to make a mighty stir among the philosophers – perhaps even among the theologians. Indeed the very reputation of such a work from such an authority as Mr Darwin would seem to have done so already, for, if we are rightly informed, the entire edition was taken off on the first day of publication. Those who have perused his 'Voyage round the World,' need not be told that the author is a man of curious and careful research, familiar with every branch of natural knowledge and gifted with the faculty of expressing himself, even on the abstrusest questions, in language always perspicuous and often eloquent.

'Mr Darwin's work, although extending to 500 pages, is but an abstract of a greater which he is preparing, and which two or three years hence will be completed. The doctrine he adopts to account for the present condition of the living world is, in fact, a revival of an old one of the transmutation of species; but he illustrates it with an amount of knowledge and ingenious appliances never before brought to its support. We are ourselves by no means convinced by his reasoning, nor do we think that it overthrows the existing theory of philosophers, founded on the evidences of geological discoveries, that the organic world, as we see it, is the result of a succession of creations and destructions. There will, however, no doubt, be many converts to Mr Darwin's opinions, which for the perfect integrity with which they are stated are entitled to the most respectful study.'

*

'A mighty stir' it most definitely did create. Vociferous opponents included the palaeontologist and soon to be founder of the Natural History Museum in London, Richard Owen. He passionately believed in the idea of 'fixed species', which had been the cornerstone of his scientific work. But this was something that the *Origin* by definition rejected. Owen's work was a major influence on the then Bishop of Oxford, Samuel Wilberforce, whose infamous 'debate' with Huxley at the 1860 meeting of the British Association contained a quip from the Bishop about man being descended from apes, which was intended to ridicule. Huxley recalled the celebrated episode, along with his reply; 'If then, said I, the question is put to me would I rather have a miserable ape for a grandfather or a man highly endowed by nature and possessed of great means of influence & yet who employs these faculties & that influence for the mere purpose of introducing ridicule into a grave scientific discussion, I unhesitatingly affirm my preference for the ape.'

From Sir Gavin Rylands De Beer's *Charles Darwin: Evolution by Natural Selection* (1963):

Even more remarkable, among the answers to questions at a recent examination for the General Certificate of Education at Advanced level was the following: 'Darwin's theory was based upon three good solid pints [sic]; 1. the struggle for exits 2. the survival of the fattest 3. maternal election.'

'THE FOUR STAGES OF PUBLIC OPINION'

T. H. Huxley was a firm believer that *magna est veritas et praevalebit* (truth is great and will prevail) but, as he drolly remarked, 'truth is great, certainly, but considering her greatness, it is curious what a long time she is apt to take about prevailing'. He expanded upon this sentiment in a notebook jotting regarding 'The Four Stages of Public Opinion'. It's hard to disagree with his analysis.

I (Just after publication)
The Novelty is absurd and subversive of Religion and Morality.
    The propounder both fool and knave.

II (Twenty years later)
The Novelty is absolute Truth and will yield a full and satisfactory explanation of things in general.
    The Propounder man of sublime genius and perfect virtue.

III (Forty years later)
The Novelty won't explain things in general after all and therefore is a wretched failure.
    The Propounder a very ordinary person advertised by a clique.

IV (A century later)
The Novelty is a mixture of truth and error. Explains as much as could reasonably be expected.
    The propounder worthy of all honour in spite of his

share of human frailties, as one who has added to the permanent positions of science.

The Swiss palaeontologist Louis Agassiz (1807–1873) made a similar observation when he said, 'Every great scientific truth goes through three stages. First people say it conflicts with the Bible. Next they say it had been discovered before. Lastly they say they always believed it'.

'ENDLESS FORMS MOST BEAUTIFUL'

The final paragraph of *On the Origin of Species* is worthy of an entry of its own since it sums up Darwin's thesis so beautifully, and succinctly points out that a scientific understanding of the world need not denude it of its wonder – it can, in fact, enhance it:

It is interesting to contemplate an entangled bank, clothed with many plants of many kinds, with birds singing on the bushes, with various insects flitting about, and with worms crawling through the damp earth, and to reflect that these elaborately constructed forms, so different from each other, and dependent on each other in so complex a manner, have all been produced by laws acting around us. These laws, taken in the largest sense, being Growth with Reproduction; Inheritance, which is almost implied by Reproduction; Variability from the indirect and direct action of the external conditions of life, and from use and disuse; a Ratio of Increase so high as to lead to a Struggle for Life, and as a consequence to Natural Selection, entailing Divergence of Character and the Extinction of less-improved forms. Thus, from the war of nature, from famine and death, the most exalted object which we are capable of conceiving, namely, the production of the higher animals, directly follows. There is grandeur in this view of life, with its several powers, having been originally breathed into a few forms or into one; and that, whilst this planet has gone cycling on according to the fixed law of gravity, from so simple a beginning endless forms most beautiful and most wonderful have been, and are being, evolved.

ARE YOU CLEVERER THAN A FIFTEEN-YEAR-OLD … FROM 1858?

A national curriculum was first introduced into the UK in 1988. Before that, a school's choice of examination board determined what a student was to learn for that subject. The first of these boards started in 1858 when both the universities of Oxford and Cambridge ran local public examinations for children aged under sixteen (junior candidates) and under eighteen (senior candidates) for the first time. The Cambridge exams were held in eight locations, including Birmingham, Grantham, Liverpool and Norwich. However, these exams were only public and egalitarian up to a point – the cost to sit one was £1, again an amount close to a policeman's weekly wage.

Set by the university's dons, the junior examinations included papers on 'pure mathematics', 'mechanics and hydrostatics', 'chemistry' (both theoretical and practical) and 'zoology and botany'. In total, there were ten sections examined and every student had to pass three while entering no more than six. Religious knowledge, the first of the sections, was compulsory unless the student's parents or guardians objected. Ironically, the examiner's first report mentioned there being 'some difficulty in settling the examination in Religious Knowledge owing to the great difference of opinion which appeared to prevail concerning it'.

Many of the mathematics questions make sense to us today, even if a few are beyond the expected knowledge of contemporary sixteen-year-olds, especially when they involve log tables. This is true also of a number of the science questions although they also serve to illustrate how much was yet to be discovered and also how scientific language has changed. Here are some sample questions from the junior paper.

*

Pure mathematics
Explain how to find the sum of a series of n terms in arithmetical progression, whose first term is a, and last term l.

Sum the series: 6, −2, ⅔, −²⁄₉, &c. to infinity.

Mechanics and Hydrostatics
Find the ratio of the power to the weight in a system of pullies where all the strings are attached to an uniform bar from which the weight is suspended, the weights of the pullies being neglected. From what point of the bar ought the weight to be suspended that the bar may rest in a horizontal position?

A is a fixed pully, *B*, *C* *heavy* moveable pullies. An inextensible string without weight is thrown over *A*. One end of it passes under *C* and is fastened to the centre of *B*, the other end passes under *B* and is fastened to the centre of *A*. Compare the weights of *B* and *C* that the system may be in equilibrium; the strings being all parallel.

Chemistry
Name the different compounds of Nitrogen and Oxygen, and state the composition of each. By what qualities could you recognize the deutoxide (or binoxide) of Nitrogen?

Zoology
Which Vertebrates are oviparous; which are abranchiate; and which have gills during only a period of their existence? Which mammals have the simplest kind of teeth?

Botany
What parts of plants are the Blackberry, Strawberry, Mulberry, Apple, Potato, Beet, Tea and Opium?

*

How did you do? I won't tell you what I got, obviously out of modesty.

BY DEGREES

This graph shows what percentage of new degrees in a range of countries were in science or engineering in 2007. It's rather sad to see that there has been a general decline.

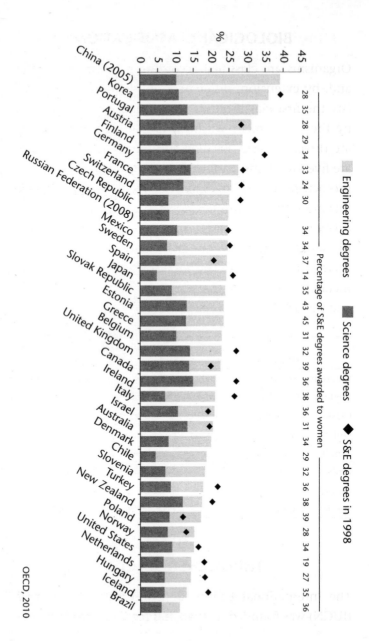

The chart shows the percentage of S&E (Science and Engineering) degrees, broken down by Engineering degrees and Science degrees, with markers for S&E degrees in 1998 and the percentage of S&E degrees awarded to women.

%

45
40
35
30
25
20
15
10
5
0

China (2005)
Korea   28
Portugal   35
Austria   28
Finland   29
Germany
France   34
Switzerland   33
Czech Republic   24
Russian Federation (2008)   30
Mexico
Sweden   34
Spain   34
Japan   37
Slovak Republic   14
Estonia   35
Greece   43
Belgium   45
United Kingdom   31
Canada   32
Ireland   39
Italy   36
Israel   38
Australia   36
Denmark   31
Chile   34
Slovenia   29
Turkey   32
New Zealand   36
Poland   38
Norway   39
United States   28
Netherlands   34
Hungary   19
Iceland   27
Brazil   35
                     36

Engineering degrees

Science degrees

◆ S&E degrees in 1998

Percentage of S&E degrees awarded to women

OECD, 2010

THE SCIENCE MAGPIE

BIOLOGICAL CLASSIFICATION

Organisms are classified according to their similarities and, by extension, differences. By definition, they satisfy the first classification, life, which is then followed by their placement within a domain, of which there are three: Archaea, Bacteria and Eucaryota. Then there are five kingdoms – animals, bacteria, fungi, plants and protoctista (e.g. sponges and seaweed). The taxonomic hierarchy then continues down through the following categories: phylum, class, order, family (sometimes with the subgroup tribe), genus and finally species, of which there exist millions. The table below shows how the hierarchy works with humans, lions, pacific bluefin tuna and the damask rose as examples.

	Humans	Lion	Pacific bluefin tuna	Damask rose
Kingdom	Animalia	Animalia	Animalia	Plantae
Phylum	Chordata	Chordata	Chordata	Embryophyta
Class	Mammalia	Mammalia	Actinopterygii	Magnoliopsida
Order	Primates	Carnivora	Perciformes	Rosales
Family	Hominidae	Felidae	Scombridae	Rosaceae
(Tribe)	Hominini		Thunnini	
Genus	Homo	Panthera	Thunnus	Rosa
Species	H. sapiens	P. leo	T. orientalis	R. damascena

THREATENED SPECIES

The International Union for Conservation of Nature (IUCN) was founded in 1948 and regularly produces the

Red List of Threatened Species, which gives information on the 'status of wild species and their links to livelihoods'. Species are categorised using a system of eight labels:

EX – Extinct
EW – Extinct in the wild
CR – Critically endangered
EN – Endangered
VU – Vulnerable
LR – Lower risk
DD – Data deficient
NE – Not evaluated

A threatened species is defined as any listed as Critically Endangered (CR), Endangered (EN) or Vulnerable (VU). The following table gives you some idea of the scope of the survey and shows the number of species currently classed as threatened.

	Estimated number of described species	Number of species evaluated by 2012	Number of threatened species in 2012
VERTEBRATES			
Mammals	5,501	5,501	1,140
Birds	10,064	10,064	1,313
Reptiles	9,547	3,663	802
Amphibians	6,771	6,370	1,931
Fishes	32,400	10,359	2,041
Subtotal	64,238	35,957	7,227

(continued)

	Estimated number of described species	Number of species evaluated by 2012	Number of threatened species in 2012
INVERTEBRATES			
Insects	1,000,000	3,899	776
Molluscs	85,000	6,028	1,729
Crustaceans	47,000	2,399	596
Corals	2,175	856	235
Arachnids	102,248	33	19
Velvet worms	165	11	9
Horseshoe crabs	4	4	0
Others	68,658	52	24
Subtotal	1,305,250	13,280	3,388
Total	**1,369,533**	**49,237**	**10,615**

In July 2012 it was reported that over 90% of the 103 species of lemur should be classed as threatened, with 23 of them Critically Endangered – fifteen more than in 2008.

THE BIG FIVE

A mass extinction is defined as the disappearance of a dramatic number of families (typically more than 10%) or species (usually more than 40%) in a relatively short space of time (in relation to the geological timescale, so in actual fact we could still be talking about hundreds of thousands of years). It's believed that there have been more than twenty such events in Earth's history but in 1982 American palaeontologists Jack Sepkoski and David M. Raup proposed a list of 'The Big Five'.

Extinction	When	Estimated size/ impact	Main theory regarding cause
Ordovician-Silurian mass extinction	Approx. 440 million years ago	86% of species	Climate cooling due to glaciation
Late Devonian mass extinction	Approx. 360 million years ago	75% of species	Changes in sea level, a decrease in oxygen concentrations in the sea, global cooling
Permian mass extinction	Approx. 250 million years ago	96% of species	Volcanism, global warming and ocean acidification
Triassic-Jurassic mass extinction	Approx. 200 million years ago	80% of species	Increase in atmospheric CO_2 leading to global warming and ocean chemistry change
Cretaceous-Tertiary mass extinction	Approx. 65 million years ago	76% of species	Extraterrestrial impact (asteroid/ meteor) leading to rapid cooling

In total, 99% of species that have ever lived on Earth no longer exist.

THE LAST 500 MILLION YEAR EVOLUTION OF VERTEBRATES

In *Vertebrate Palaeontology*, Professor Mike Benton of the University of Bristol includes a spindle diagram showing 'the pattern of evolution of the vertebrates'. With his kind permission it is reproduced here.

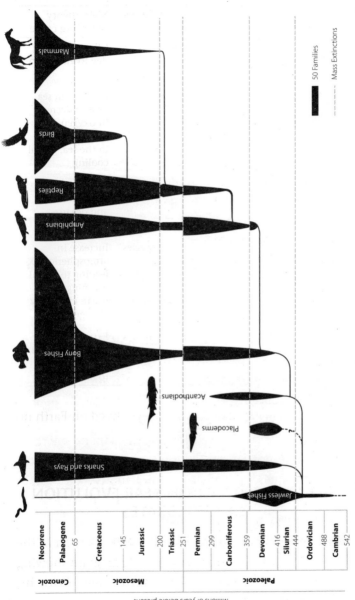

Millions of years before present

Cenozoic	Neogene			
	Palaeogene	65		
Mesozoic	Cretaceous	145		
	Jurassic	200		
	Triassic	251		
Paleozoic	Permian	299		
	Carboniferous	359		
	Devonian	416		
	Silurian	444		
	Ordovician	488		
	Cambrian	542		

Mammals
Birds
Reptiles
Amphibians
Bony Fishes
Acanthodians
Placoderms
Sharks and Rays
Jawless Fishes

■ 50 Families

- - - Mass Extinctions

THE TORINO IMPACT HAZARD SCALE

In the 1979 disaster movie *Meteor*, a 5-mile wide asteroid is on a collision course with the Earth. The planet's only hope is a collaboration between the USA and Soviet Union involving their respective orbiting nuclear missile launchers. It's fair to say that the film isn't the finest hour for its all-star cast (which includes Sean Connery, Natalie Wood, Karl Malden, Trevor Howard, Martin Landau and Henry Fonda) but it does serve to highlight the potential of disaster from outer space (the K-T event that wiped out the dinosaurs and much more perhaps serves as a more highbrow example).

In 1979, there was no formalised way of expressing the seriousness of such a situation. It would be another sixteen years before, in 1995, Professor Richard P. Binzel first presented the then unnamed Torino impact hazard scale to provide an effective way of assessing the potential effect upon the Earth of any newly discovered asteroid or comet and then communicating that to the general public. Expect to find it cropping up in some disaster movie or other sooner or later.

The scale works by plotting the kinetic energy of the extraterrestrial object (mv^2 i.e. mass × its expected velocity squared) against the probability of it striking the Earth. This value isn't necessarily fixed – it can change over time as information regarding the object improves.

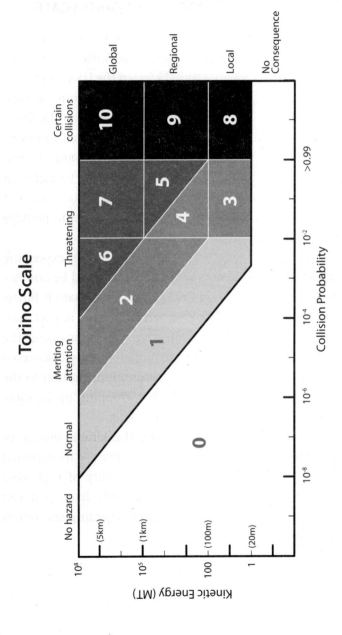

No hazard (White Zone)	0	The likelihood of a collision is zero, or is so low as to be effectively zero. Also applies to small objects such as meteors and bodies that burn up in the atmosphere as well as infrequent meteorite falls that rarely cause damage.
Normal (Green Zone)	1	A routine discovery in which a pass near the Earth is predicted that poses no unusual level of danger. Current calculations show the chance of collision is extremely unlikely with no cause for public attention or public concern. New telescopic observations very likely will lead to re-assignment to level 0.
Meriting attention by astronomers (Yellow Zone)	2	A discovery, which may become routine with expanded searches, of an object making a somewhat close but not highly unusual pass near the Earth. While meriting attention by astronomers, there is no cause for public attention or public concern as an actual collision is very unlikely. New telescopic observations very likely will lead to re-assignment to level 0.
Meriting attention by astronomers (Yellow Zone)	3	A close encounter, meriting attention by astronomers. Current calculations give a 1% or greater chance of collision capable of localised destruction. Most likely, new telescopic observations will lead to re-assignment to level 0. Attention by public and by public officials is merited if the encounter is less than a decade away.

(continued)

THE SCIENCE MAGPIE

Meriting attention by astronomers (Yellow Zone)	4	A close encounter, meriting attention by astronomers. Current calculations give a 1% or greater chance of collision capable of regional devastation. Most likely, new telescopic observations will lead to re-assignment to level 0. Attention by public and by public officials is merited if the encounter is less than a decade away.
Threatening (Orange Zone)	5	A close encounter posing a serious, but still uncertain threat of regional devastation. Critical attention by astronomers is needed to determine conclusively whether or not a collision will occur. If the encounter is less than a decade away, governmental contingency planning may be warranted.
Threatening (Orange Zone)	6	A close encounter by a large object posing a serious but still uncertain threat of a global catastrophe. Critical attention by astronomers is needed to determine conclusively whether or not a collision will occur. If the encounter is less than three decades away, governmental contingency planning may be warranted.
Threatening (Orange Zone)	7	A very close encounter by a large object, which if occurring this century, poses an unprecedented but still uncertain threat of a global catastrophe. For such a threat in this century, international contingency planning is warranted, especially to determine urgently and conclusively whether or not a collision will occur.

(continued)

Certain collision (Red Zone)	8	A collision is certain, capable of causing localised destruction for an impact over land or possibly a tsunami if close offshore. Such events occur on average between once per 50 years and once per several 1,000 years.
Certain collision (Red Zone)	9	A collision is certain, capable of causing unprecedented regional devastation for a land impact or the threat of a major tsunami for an ocean impact. Such events occur on average between once per 10,000 years and once per 100,000 years.
Certain collision (Red Zone)	10	A collision is certain, capable of causing global climatic catastrophe that may threaten the future of civilisation as we know it, whether impacting land or ocean. Such events occur on average once per 100,000 years, or less often.

Copyright ©1999, 2004 Richard P. Binzel, Massachusetts Institute of Technology.

At the time of writing, there were two objects with a Torino scale value of 1 and nothing higher, so there's no need to panic.

> The great tragedy of science – the slaying of a beautiful hypothesis by an ugly fact.
>
> T. H. Huxley. 'Biogenesis and Abiogenesis'.
> Collected Essays viii

THE $10 MILLION BOOK

At 5pm on 7 December 2010, the London-based auctioneer Sotheby's began a session consisting of 91 lots entitled 'Magnificent Books, Manuscripts and Drawings from the Collection of Frederick 2nd Lord Hesketh'. These included a first folio edition of Shakespeare's *Comedies, Histories & Tragedies* from 1623, and Higden's *Polychronicon* printed by William Caxton in 1482. But the undoubted star of the collection was a first edition, comprising four volumes, of *The Birds of America*, which had been published by the author, John James Audubon, between 1827 and 1838. It sold for $10.3 million and became the most expensive book ever bought at auction (the previous holder of the record was also *The Birds of America* and, reporting on the sale, *The Economist* revealed that 'a list of the ten most-expensive books would include five copies of *The Birds of America*').

There are only 120 copies of *The Birds of America* still known to exist. Measuring roughly 97cm × 65cm, the book is a collection of 435 magnificent hand-coloured etched plates, with line-engraving and aquatint, detailing many of the bird species of the United States including some now extinct. Its production costs were so high that it was published in 87 sets of five plates over an eleven year period. It is, quite simply, a work of art. It is also an ornithological masterpiece and had a profound influence on many natural historians, including Charles Darwin, who would go on to refer to Audubon three times in his *On the Origin of Species*.

Below is a list, based on the *Economist*'s article, of the six most expensive science books bought at auction.

Title	Author	Year edition published	Value
The Birds of America	John James Audubon (1785–1851)	1827–1838	$10.3 million (2010)
Les Liliacées	Pierre-Joseph Redouté (1759–1840)	1802	$5 million (1985)
Geographia – Cosmographia	Claudius Ptolemy (c. AD 90–c. AD 168)	1462 edition	$3.5 million (2006)
De revolutionibus orbium coelestium	Nicolaus Copernicus (1473–1543)	1543	$1.9 million (2008)
Complete Folio of *Birds*	John Gould (1804–1881)	1831–1888	$1.8 million (1998)
De humani corporis fabrica	Andreas Vesalius (1514–1564)	1543	$1.5 million (1998)

THE POETS' SCIENTIST

Aged not quite twenty and already showing significant scientific promise, Cornishman Humphry Davy joined the Pneumatic Institution in Bristol in 1798. During the following year he investigated the effects of nitrous oxide, or laughing gas, which Joseph Priestley had discovered more than twenty years earlier. The inventor James Watt built Davy a portable gas chamber and was one of the 60 'breathers' who took part in the studies, along with Davy's good friends the poets Samuel Taylor Coleridge and Robert Southey. As well as recording and publishing the results of his experiments, Davy penned the following lines to describe the gas's effects, which he seems to have rather enjoyed:

Not in the ideal dreams of wild desire
Have I beheld a rapture-wakening form:
My bosom burns with no unhallow'd fire,
Yet is my cheek with rosy blushes warm;
Yet are my eyes with sparkling lustre fill'd;
Yet is my mouth replete with murmuring sound;
Yet are my limbs with inward transports fill'd,
And clad with new-born mightiness around.

Poetry was to prove a life-long love for Davy. His poems were published in an anthology in 1799, including one he wrote aged just seventeen, and Davy was asked by William Wordsworth to proofread the second edition of *Lyrical Ballads* published the following year. Coleridge would later write that Davy was 'the Man who born first a poet first converted Poetry into Science'.

As a scientist, Davy built on the work of Antoine Lavoisier whose book *Traité élémentaire de chimie*, published in 1789, was considered the first true chemistry book. As well as inventing his famous lamp, which allowed miners to work much more safely by dramatically reducing the risk of underground explosions, Davy developed the method of electrolysis, using this to isolate elements such as potassium and calcium for the first time. He also gave chlorine its name. He was an acclaimed populariser of science, his lectures selling out. He succeeded Joseph Banks as president of the Royal Society in 1820 and is reported to have humbly said that his greatest discovery was his assistant Michael Faraday, who would go on to eclipse him in the annals of science. Ultimately, the lines of the last stanza of 'The Sons of Genius', the anthologised poem he wrote when seventeen, proved accurate of both himself and Faraday:

Theirs is the glory of a lasting name
The meed of Genius and her living fires,
Theirs is the laurel of eternal flame,
And theirs the sweetness of the Muse's lyres.

WHEN LEFT CAN BE RIGHT AND RIGHT CAN BE WRONG

And now for something about mirrors and molecules. It's a curious fact that the molecules chiefly responsible for the respective smells of spearmint and caraway seeds are exactly the same, except for one crucial difference.

Spearmint   Caraway

They're mirror images of the same molecule, carvone. This feature in chemistry is known as chirality, from the

Greek word *cheir* meaning hands. Your hands are chiral. Your left hand is the mirror image of your right but you can't superimpose one upon the other. So, in relation to carvone we might say that the right-hand version is responsible for the smell of caraway seeds and the left-hand version is responsible for that of spearmint. This rather implies that our smell receptors are chiral too, otherwise we wouldn't be able to distinguish between the two molecules and caraway and spearmint would therefore smell the same. Think of each molecule as a yale key and the other as its mirror image. Both keys won't fit the same lock.

It was Louis Pasteur who first noticed this feature when, using just a pair of tweezers, he painstakingly separated two different forms of tartaric acid crystals – left- and right-hand versions – and discovered they each rotated polarised light to the same degree but in opposite directions. A 50:50 mixture produced no effect at all, the two cancelling each other out.

Chirality crops up in all sorts of areas of nature, including drug synthesis. When a drug such as ibuprofen is made it typically comprises a mixture of left- and right-hand versions, called a racemic mixture, which can have important consequences. It's highly likely that only one of these two forms will do the job it's intended to do. The ideal scenario would be that the other version has no effect whatsoever and is eventually excreted by the body. But at worst, it can have adverse side effects. This is exactly what happened with the drug thalidomide in the late 1950s. It was administered as an anti-nausea drug to pregnant women in its racemic form. Disastrously, the left-hand version of the molecule caused defects and deformities in fetuses growing in the womb. Drug companies now pay a significant amount of attention to the effects different chiral versions of the drug they manufacture produce, spending vast amounts of money on producing versions that include the drug in only its pure right or left-hand form.

THE SCIENTIFIC EQUIVALENT OF: HAVE YOU READ A WORK OF SHAKESPEARE'S?

The novelist C.P. Snow, a former research scientist and civil servant, gave the annual Rede lecture at the University of Cambridge in 1959. Entitled, 'The Two Cultures and the Scientific Revolution', it not only gave birth to a book, *The Two Cultures*, that has never been out of print since, it also launched a public debate that has perhaps never been satisfactorily settled – how the arts (or humanities) and science can best work with and support each other, particularly with regard to directing public and government policy.

At one point in the book, Snow describes occasions when he has been in the company of 'highly educated' people scathing of scientists who have 'never read a major work of English literature'. Occasionally, Snow felt the need to respond by asking how many of them could describe the second law of thermodynamics? According to Snow, their subsequent incredulity showed that they failed to see his point that this was the equal of their disparagement – it was, he felt, 'the scientific equivalent of: have you read a work of Shakespeare's?'.

Snow was not merely sniping back at the detractors of science; his overall point was, to borrow a phrase from Shakespeare, that these two cultures, and society at large, would benefit most by being 'two distincts, division none'.

The next entry in the book will, amongst other things, describe the second law of thermodynamics. Since there was not room to include a full work by the Bard within this book you will have to take the initiative and go away and read one yourself, if you haven't already. After that you will, it is to be hoped, be a fully rounded individual, ready to take your place as a valued member of society.

YOU CAN'T WIN, YOU KNOW

In his *Autobiographical Notes*, Albert Einstein argued that 'a theory is the more impressive the greater the simplicity of its premises, the more different kinds of things it relates, and the more extended its area of applicability'. The example he gave of a theory demonstrating these qualities was classical thermodynamics, which Einstein was 'convinced … would never be overthrown'.

The industrial revolution and creation of the heat engine kick-started the concerted study of thermodynamics – a word not coined until William Thomson, later Lord Kelvin, put it forward in an 1843 paper. Within ten years of this, versions of what would become the first and second laws of thermodynamics had been proposed, both framed within a classical view of thermodynamics, meaning they were based on what was empirically measurable, typically within a laboratory setting.

The 1870s and beyond saw the development of statistical thermodynamics due to work done by scientists such as James Clerk Maxwell and Ludwig Boltzmann. Here probability theory was employed to understand the laws at the microscopic level of atoms and molecules. Around the same time, chemical thermodynamics, relating to chemical reactions and states of matter, was born. These two branches of thermodynamics would eventually lead to the proposal of the third law in 1912 by chemist Walter Nernst.

The zeroth law was the last to be formally recognised, although the primacy of its name is testament to the fact it had been stated in various guises before and is something fundamental to the other three laws.

Each of the four laws can be stated in a variety of ways, typically depending on the branch of thermodynamics and the context in which it is being described: classical, statistical, chemical etc. The versions that follow are essentially the form in which they were first stated:

The zeroth law
If two systems are in thermal equilibrium with a third system, they are also in thermal equilibrium with each other.
Which means all three systems would be at the same temperature. It is this law that allows a thermometer to

do its job – it falls into thermal equilibrium with the system it's measuring and we can read the temperature. If a third system is the same temperature, then all three are in thermal equilibrium with each other. The zeroth law also says that the net heat flow between two objects at the same temperature is zero.

The first law
Energy can be neither created nor destroyed, only interconverted between forms.

This law is in effect based on the principle of conservation of energy. For example, if you walk up a snowy mountain carrying a pair of skis, some of the *chemical energy* stored in your body is converted into *potential energy* (and almost certainly *heat* – make sure you wear layers), which is, in essence, a store of energy. If you now put your skis on and journey back down, this *potential energy* is converted into *kinetic energy* (this is energy related to motion). Other forms of energy include electrical, light, sound and, as Albert Einstein showed with his famous equation $E = mc^2$, mass.

The second law
Heat energy always flows spontaneously from a hot to a cold system and never naturally the other way round.

If you place ice in warm water, heat flows into it from the surrounding water. As the word 'naturally' in the above implies, heat *can* flow from cold to hot but to do this, energy is needed. If you think about it, this is exactly how ice is made in a freezer – the required energy is supplied through electricity (electrical energy). One implication of the second law is that it's impossible to make a perfect heat engine in which all the heat generated, such as in a combustion engine, can be converted into work – some of the heat *must* be lost to any surrounding colder system.

The third law

A system can't be reduced to absolute zero (0K) in a finite number of steps.

This is probably the hardest of the laws to grasp and the least relevant to everyday life. Which probably explains why it wasn't chosen by C. P. Snow to bolster his argument.

> *If someone points out to you that your pet theory of the universe is in disagreement with Maxwell's equations – then so much the worse for Maxwell's equations. If it is found to be contradicted by observation – well, these experimentalists do bungle things sometimes. But if your theory is found to be against the second law of thermodynamics I can give you no hope; there is nothing for it but to collapse in deepest humiliation.*
>
> Arthur Eddington, *The Nature of the Physical World* (1928)

THE MPEMBA EFFECT

First revealed in *Physics Education* in 1969, this is a story to lift the heart, and is a powerful lesson to us all. In

1963, Erasto B. Mpemba was in his third year at secondary school in Tanzania. During a physics lesson, he asked his teacher about something that had been puzzling him – why was it that the ice-cream mixture he'd made which had been boiling had frozen quicker than his friend's more tepid mixture, even though they had been placed in the freezer at the same time. He was simply told he must have been confused.

Mpemba didn't just accept this brush off and the problem continued to trouble him. His conviction in the truth of his observations increased as friends who regularly made ice-cream confirmed it was quicker to do so the hotter the mixture was when put in the freezer.

After passing his O level a few years later, Mpemba found himself in high school. The first topic he studied there was heat. He asked his new teacher the 'hotter ice-cream freezing faster' question and again he was told he was confused and even guilty of believing in 'Mpemba physics' as opposed to real physics. The teacher didn't let this go and it became a running joke with both him and Mpemba's classmates. In spite of this, Mpemba persisted and experimented further with water in beakers. His findings proved the same as before.

One day Dr D. G. Osborne of University College Dar es Salaam visited the school and, at the end of the session, invited questions from the students. Mpemba bravely asked, 'If you take two similar containers with equal volumes of water, one at 35°C and the other at 100°C, and put them into a refrigerator, the one that started at 100°C freezes first. Why?' The rest of the audience sniggered but Dr Osborne asked Mpemba to confirm he'd performed the experiment before promising to try it himself, even though he thought him mistaken. Dr Osborne recognised the 'need to encourage students to develop questioning and critical attitudes' and felt that that there could be a 'danger… [in] authoritarian physics'.

Much to his surprise, Dr Osborne got the same results, as did the university students he put to work on the problem. Furthermore, no answer could be found in any scientific literature.

Finally, in the paper in *Physics Education*, jointly authored by Mpemba and Dr Osborne, a request was made to readers for further information. It turned out that the question had already puzzled many acclaimed minds, including Aristotle, Francis Bacon and René Descartes. Today, the problem is fittingly known as the Mpemba Effect and while many explanations have been put forward none have been proven – the jury is still out. Towards the end of June 2012, the *Royal Society of Chemistry* began a competition to find the 'best and most creative explanation' for the effect. Whether the puzzle is solved soon or not, we must all admire Mpemba for his tenacity and self-belief.

MAXWELL'S DEMON

On 11 December 1867, James Clerk Maxwell wrote a letter to his close friend the Scottish physicist and mathematician Peter Guthrie Tait. In it he described a thought experiment that violated the second law of thermodynamics. In Maxwell's eyes, it was designed to show that the law was guaranteed only in a statistical sense. This is how he introduced the experiment in a book published four years later.

'One of the best established facts in thermodynamics is that it is impossible in a system enclosed in an envelope which permits neither change of volume nor passage of heat, and in which both the temperature and the pressure are everywhere the same, to produce an inequality of temperature or pressure without the expenditure of work. This is the second law of thermodynamics and it is undoubtedly true as long as we can deal with bodies only in mass and have no power of perceiving or handling the separate molecules of which they are made up.'

But in his thought experiment Maxwell imagines 'a being whose faculties are so sharpened that he can follow every molecule in its course, such a being whose attributes are still essentially finite as our own, would be able to do what is at present impossible for us'. This is the 'demon'. He continued:

'We have seen that the molecules in a vessel full of air at uniform temperature are moving with velocities by no means uniform, though the mean velocity of any great number of them, arbitrarily selected, is almost exactly uniform.

'Now let us suppose that such a vessel is divided into two portions, A and B, by a division in which there is a small hole, and that a being, who can see the individual molecules, opens and closes this hole, so as to allow only the swifter molecules to pass from A to B, and only the slower ones to pass from B to A. He will thus, without expenditure of work, raise the temperature of B and lower that of A in contradiction to the second law of thermodynamics.'

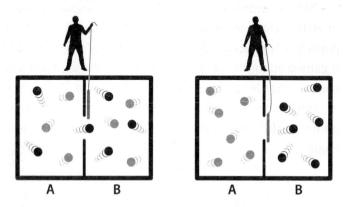

Writing again about his idea in a letter to John William Strutt, Maxwell summarised the significance of his idea as follows: 'The second law of thermodynamics has the same degree of truth as the statement that if you throw a tumblerful of water into the sea you cannot get the same tumblerful of water out again.' That is to say, although in practical, real-life terms it is true, it is not *absolutely* true.

In Maxwell's scenario, the demon is using only information to produce a temperature difference – and therefore the necessary ingredients to be able to run a heat engine (a steam engine is a good example of one of these) – from a system in thermal equilibrium. In 1929, the Hungarian physicist Leó Szilárd (who would four years later conceive the idea of the nuclear chain reaction) argued that the demon's processing of information on each molecule's temperature *had* to involve energy, more than was subsequently gained, meaning that the second law wasn't in fact violated. The implication, though, was that information and energy were related.

Rather astonishingly, in the journal *Nature Physics* in 2010 a team of Japanese scientists reported making a particle 'climb up a spiral staircase-like potential', that is it gained energy, through the use of a real-time feedback control, thereby gaining more energy than had been put in. While there is still a long way to go, their research points to the incredible idea of an 'information-to-heat engine'.

HOTTER THAN HELL

As any good Catholic will tell you, eternal damnation ain't what it used to be. This may partly be due to this wonderful analysis that appeared in the journal *Applied Optics* in 1972. It shows that, contrary to expectation, Heaven is in fact hotter than Hell.

*

The temperature of Heaven can be rather accurately computed. Our authority is the Bible. Isaiah 30:26 reads:

Moreover, the light of the Moon shall be as the light of the Sun and the light of the Sun shall be sevenfold as the light of seven days.

Thus, Heaven receives from the Moon as much radiation

as the Earth does from the Sun, and in addition seven times seven (49) times as much as the Earth does from the Sun, or 50 times in all. The light we receive from the Moon is one ten-thousandth of the light we receive from the Sun, so we can ignore that. With these data we can compute the temperature of Heaven: the radiation falling on Heaven will heat it to the point where the heat lost by radiation is just equal to the heat received by radiation. In other words, Heaven loses 50 times as much heat as the Earth by radiation. Using the Stefan-Boltzmann fourth power law for radiation

$$(H/E)^4 = 50$$

where E is the absolute temperature of the earth, 300°K (273+27). This gives H, the absolute temperature of heaven, as 798°K absolute (525°C).

The exact temperature of Hell cannot be computed but it must be less than 444.6°C, the temperature at which brimstone or sulfur changes from a liquid to a gas. Revelations 21:8: *But the fearful and unbelieving ... shall have their part in the lake which burneth with fire and brimstone.* A lake of molten brimstone [sulfur] means that its temperature must be at or below boiling point, which is 444.6°C. (Above that point, it would be a vapour, not a lake.)

We have then, temperature of Heaven, 525°C. Temperature of Hell, less than 445°C. Therefore Heaven is hotter than Hell.

*

A letter to the journal *Physics Today* in 1998 from two scientists at the University of Santiago pointed out a potential flaw in one of the calculations. They argued that the Isaiah passage had been misread and that the radiation from the Sun received by Heaven is merely seven times what the Earth receives and not 49 times. If this is the case we must revise the temperature in Heaven to 231.5°C. Still unpleasantly hot, but a lot of people felt relieved that a sense of Biblical balance had been restored.

ON THE OTHER SIDE OF SILENCE

In July 2009, the heavy metal band Kiss reportedly achieved 136 decibels during a live performance in Canada, thereby making them one of the loudest bands in the world. As we will see in a moment, that should be beyond the threshold of pain, although many would argue Kiss already surpass that the moment they step on stage.

We measure loudness in decibels. A decibel is a unit used to quantify the ratio between two power levels, one of which is treated as the reference level. In the case of measuring loudness, this reference level is taken to be the threshold of human hearing. The (rather complicated-looking) formula for its calculation is:

$$\text{Number of decibels} = 10 \ \log_{10}\left(\frac{P}{P_0}\right)$$

where P is the level effectively being measured and P_0 is the reference. As such, a difference in loudness of 10 decibels, say from 30dB to 40dB, means that one sound is ten times louder than the other. A change of one decibel represents an increase in loudness of approximately 26%. This gives a context to the figures you often hear about.

dB	Sensation
0	Threshold of hearing
10	Breathing
20	Rustling leaves
30	Watch ticking 1 metre away
40	Birdsong
50	Quiet conversation
70	Loud conversation
80	Door slamming
110	Pneumatic drill
130	Threshold of pain

Decibels are also used as units in optics and electronics, as well as other disciplines because they are so useful, but when you hear them mentioned it will almost certainly be in relation to how loud something is.

'THE CHEMICAL HISTORY OF THE CANDLE'

The son of a journeyman blacksmith, Michael Faraday would go on to become one of the most celebrated scientists of the 19th century, particularly because of his work on electricity and magnetism, including the discovery that electricity could be produced by repeatedly moving a magnet through a coil of wire. Over 50 years after Faraday's death, the novelist Aldous Huxley felt compelled to say, 'Even if I could be Shakespeare, I think I should still choose to be Faraday.' That is quite a compliment.

Faraday came to Humphry Davy's attention when, aged just 21 and having just completed his bookbinding apprenticeship, he presented Davy with a bound copy of the lecture notes he'd taken down when attending the great man's demonstrations at the Royal Institution. It was a brave move from a young man who had previously declared it 'impossible … to follow [Davy]. I should merely injure and destroy the beautiful and sublime observations that fell from his lips'. Davy apparently felt otherwise, recommending Faraday for a position as a laboratory assistant.

Faraday proved a quick learner. Within less than a year, from having felt he could never hope to be a speaker of Davy's quality, he was writing to his friend Benjamin Abbott, confidently describing what the necessary qualities of a good lecturer were. In 1825 Faraday would originate the Institution's now celebrated series of Christmas Lectures, going on to present nineteen years of them himself. One of these was called 'The Chemical History of the Candle' and was eventually published as a book in 1861, proving hugely successful. It is still available to buy today (I recommend it). Over the course of six lectures, Faraday took the audience through the remarkable science behind an item everyone had everyday knowledge of. What follows is just a brief taste of its magic and something you may want to try at home.

There is [a] condition which you must learn as regards the candle, without which you would not be able fully to understand the philosophy of it, and that is the vaporous condition of the fuel [the wax]. In order that you may understand that, let me shew you a very pretty, but very common-place experiment. If you blow a candle out cleverly, you will see the vapour rise from it. You have, I know, often smelt the vapour of a blown-out candle – and a very bad smell it is; but if you blow it out cleverly, you will be able to see pretty well the vapour into which this solid matter is transformed. I will blow out one of these candles in such a way as not to disturb the air around it, by the continuing action of my breath; and now, if I hold a lighted taper two or three inches from the wick, you will observe a train of fire going through the air till it reaches the candle. I am obliged to be quick and ready, because, if I allow the vapour time to cool, it becomes condensed into a liquid or solid, or the stream of combustible matter gets disturbed.

WHEN IT COMES TO WHAT'S IN YOUR GENES, SIZE ISN'T EVERYTHING

An organism's genome refers to all of the genes – sections of DNA involved in determining a particular physical or biological trait. These are contained in a single set of its chromosomes, pairs of rod-like structures found inside every developed cell in its body and which carry its DNA. From the inception of life this codified information determines the way an organism grows and what it becomes. This being the case one would tend to assume that the more complicated an organism is, the more genes it must have. But when genomes started to be mapped, scientists made the rather startling discovery that this is not the case. Humans, for example, were originally estimated to have about 100,000 genes but the true figure, which is still being refined, appears to be less than a quarter of this. By comparison, the genome of a particular species of rice contains over twice as many genes as a human. The following information taken from a 2008 paper in the journal *Nature Education* shows how the number of genes varies surprisingly between different organisms:

Organism	Number of genes
Trichomonas vaginalis (a single-celled parasitic organism responsible for over 160 million urinary tract infections each year)	60,000
Oryza sativa (a common rice that feeds over half the world's population)	51,000
Mus musculus (mouse of choice for pet lovers and laboratory leaders)	30,000
Homo sapiens (humans)	20,000–25,000
Drosophila melanogaster (the common fruit fly, a staple of school experiments)	14,000

(continued)

Organism	Number of genes
Saccharomyces cerevisiae (a yeast commonly used in baking and brewing)	6,000
Escherichia coli (commonly referred to as *E. coli*, this bacteria can sometimes cause food poisoning)	4,500

> *The scientist is not a person who gives the right answers, he is one who asks the right questions.*
>
> Claude Levi-Strauss, French anthropologist

WOULD YOU ADAM AND EVE IT?

You've got some alien DNA in your body. Not alien in the sense of UFOs, but stuff that wasn't originally a part of you (or rather of your ancestors way back). Mitochondria are found in pretty much every complex cell in the body. These very small organelles have their own DNA (called, funnily enough, mitochondrial DNA). It's a legacy of the fact that mitochondria were once independent bacteria that evolved to live inside larger cells such as our own. So, although they are now an integral part of the way our bodies work – producing adenosine triphosphate (ATP), the currency of energy in all living organisms – technically once upon a time they were just hitchhikers.

As if that weren't fascinating enough, one feature of this mitochondrial DNA is that it is only inherited through the female line. This makes it possible for scientists to trace our maternal lineage and incredibly such analysis has pointed to the fact that we *all* share a single common ancestor who lived nearly 200,000 years ago. That's one female to whom every single one of us is related! Unsurprisingly, she is often referred to as 'mitochondrial Eve' or 'African Eve'.

Amazingly, there is also a male equivalent of mitochondrial Eve – Y-chromosome Adam. This is because the Y-chromosome is exclusive to males and is passed from father to son. As such, it has been possible for researchers to discover that all men are descended from a single male, again from Africa. As with mEve (as her friends call her), where and even when exactly Y-Adam lived hasn't yet been fixed upon, although current studies suggest Y-Adam was some 60,000 years younger than mEve. The original toyboy then.

> A bloke walks into a pub, and asks for a pint of Adenosine Triphosphate. The barman says 'That's 80p!'

EUCLID'S ALGORITHM

Al-Khwarizmi was an Arab mathematician and the originator of algebra (from the Arabic *al-jebr*, meaning 'the reunion of broken parts'. If you can recall long, painful hours spent struggling to balance equations in school you might not feel particularly well disposed towards him, but he really did have a brilliant mind. He is reported to have once declared that 'with my two algorithms, one can solve all problems – without error, if God will!'

Algorithms are very, very useful. A series of algorithms can be used to solve the Rubik's Cube (you'll find lots of examples on the internet), the sales rankings on Amazon are established using algorithms and they're also used an awful lot in the world of finance. Needless to say, they can be very complex indeed. As such, their increasing use and influence has happened in tandem with the development of the modern computer.

The principle of an algorithm can be easily demonstrated using what's known as Euclid's algorithm, which enables us to find the greatest common factor of two numbers by using a particular sequence of divisions (a factor is a number that can divide a specified number exactly, for example, 2 is a factor of 6 because it divides into it exactly 3 times. A common factor is a number that can divide two different numbers exactly, so 4 is a common factor of 12 and 28).

First of all, we need two numbers, say 192 and 42. We then divide the larger number (192) by the smaller (42) in the following way to give a multiple of the smaller number plus a remainder:

$$192 = (4 \times 42) + 24$$

We then divide 42 by the remainder, 24.

$$42 = (1 \times 24) + 18$$

Then 24 by the remainder, 18.

$$24 = (1 \times 18) + 6$$
$$18 = (3 \times 6) + 0$$

Which tells us that the greatest common factor of 192 and 42 is 6. Pretty nifty, eh?

THE PERIODIC SNAIL

Everything on Earth is made from one or more of the 118 chemical elements shown in the periodic table. The version of the table that we use today was first proposed in 1869 by the Russian chemist Dimitri Mendeleev, who wanted to illustrate recurring trends in the properties of the different elements. To do this he put them into a systematic order. This is what makes the periodic table such a powerful tool. Knowing this order allows us to predict how elements will behave, and how they will interact with each other or possibly combine to make new substances.

In Mendeleev's time, only about 60 elements were known to science. He left spaces in his table predicting new elements, and as these and others have been discovered, the layout of the table has been refined and extended. Of the known elements, 94 occur naturally on Earth. The rest have been created by man-made nuclear reactions. This is the table you probably remember from school.

As with great works of literature, the periodic table can be appreciated on many levels, from when it's simply a useful system of categorisation during our school years to its developing subtlety and nuance as our

The periodic table

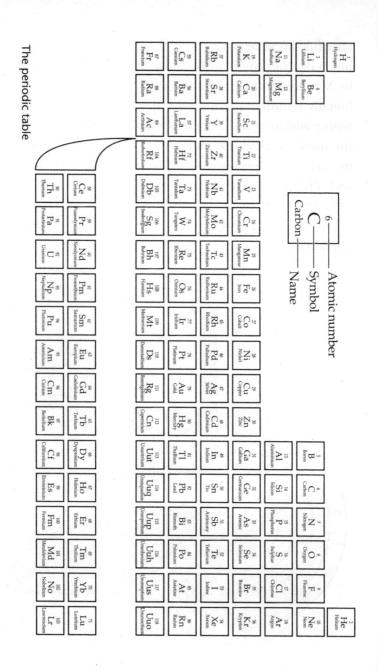

scientific knowledge grows. So it might not come as too much of a surprise to learn that ever since it was created by Mendeleev the periodic table has been presented in literally hundreds of forms; not just the one above. One theme within many of these alternative tables has been the use of a spiral. Examples include Edgar Longman's mural from the 1951 Festival of Britain Science Exhibition and Philip Stewart's Chemical Galaxy II, which is presented with an interstellar look and can be seen at www.chemicalgalaxy.co.uk. One of the best, however, is Theodore Benfey's 'periodic snail', which he developed

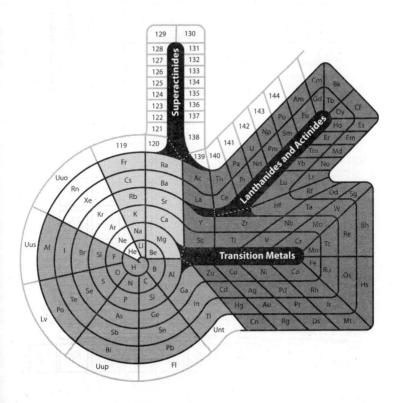

'to emphasize the complex yet beautiful periodicity in the properties of the chemical elements' and is redrawn courtesy of Theodore Benfey himself.

> The moment a bar of gold walked into
> a pub, the landlord shouted 'A U, get out!'

MAKING THE STANDARD MODEL

Okay, you might have to take a deep breath for this one – I certainly did.

Do you remember being taught that atoms are made up of three sub-atomic particles: protons, neutrons and electrons? The nucleus of an atom, found at its centre, is made up of the neutrons and protons, and only these contribute to the element's atomic mass. Both neutrons and protons are roughly the same mass. The electrons, on the other hand, 'orbit' the nucleus, and have negligible mass in comparison to it. But this description only scratches the surface.

Particle physics can be really said to have started in 1897 when J. J. Thomson discovered the electron. As this area progressed over the following years, it was often the case that theory preceded discovery. Wolfgang Pauli proposed the existence of the neutrino in 1930 but it wasn't until 1956 that it was formally detected. By the early 1960s there was a veritable 'particle zoo', and it was believed that there existed a plethora of elementary particles.

In 1964, the physicists Murray Gell-Mann and George Zweig proposed, independently of each other, a classification scheme, which became known as the quark model (Murray Gell-Mann had assigned the name quark the previous year, taking inspiration for its spelling from a dream the chief protagonist in James Joyce's *Finnegans Wake* has, in which a drunken seagull says 'three quarks for Muster Mark' – a more sober seagull would've asked for three quarts).

In the early 1970s, scientists came together to develop what has become known as the Standard Model of particles and forces. It proposes that everything in the universe is made from twelve fundamental particles and four fundamental forces (*technically*, one of these forces, gravity, isn't in the standard model as we'll soon see).

The model consists of three groups: quarks, leptons and gauge bosons, as well as the Higgs boson (we'll come to the latter in a bit). Quarks and leptons are what make up matter and are classed together as fermions. Within their respective groups, quarks and leptons are put in pairs, which are also known as generations. The 1st generation makes up stable matter.

Quarks come in six 'flavours': up (u), down (d), charm (c), strange (s), top (t) and bottom (b). Corresponding with each quark is an antiquark: $\bar{u}, \bar{d}, \bar{c}, \bar{s}, \bar{t}$ and $\bar{b}$.

When quarks are grouped in pairs, mesons are formed; when grouped in triplets they form baryons. Protons and neutrons, which form the nucleus of an atom, are types of baryon. In turn both baryons and mesons belong to the hadron family, which also includes antibaryons.

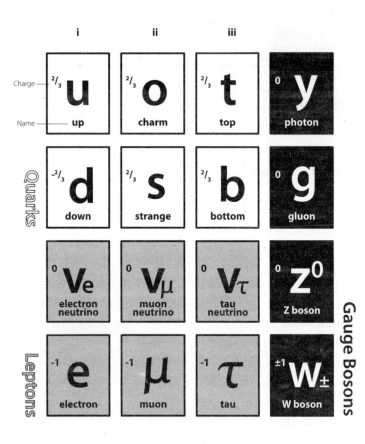

To put it another way, hadrons are all composed of different combinations of quarks which are classed as either mesons, baryons or antibaryons. Baryons consist of three quarks, antibaryons three antiquarks and mesons a quark and an antiquark.

Protons are baryons which consist of u, u and d quarks. Neutrons are also a type of baryon and comprise u, d, and d quarks. High-energy collisions can create other hadrons beside protons and neutrons. This

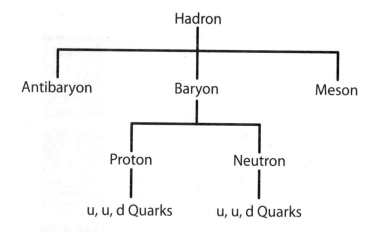

happens, for example, in the upper atmosphere where cosmic rays (high-energy protons from outer-space) collide with nuclei of oxygen and nitrogen to create a particle shower that includes mesons.

Electrons are examples of leptons, of which there are six flavours: electron (e^-), electron neutrino (v_e), muon (u^-), muon neutrino (v_u), tauon (t^-) and tauon neutrino (v_t). The electron, muon and tauon all have a charge of -1 and differ only in mass (electrons are smaller than muons which are, in turn, smaller than tauons). Neutrinos have no charge.

For each lepton there is a corresponding antilepton, which has a charge opposite to that of the matching lepton.

So there you have it, a somewhat exhausting whistle stop tour of the particle zoo, including the 12 elementary particles from which all matter in the universe is composed. Now let's take a look at the four fundamental forces of nature while you still have the will to live.

The four fundamental interactions of nature and gauge bosons

The strong force – This is the force between quarks that allows the formation of protons and neutrons. It also acts between protons and neutrons themselves and is responsible for giving the nuclei of atoms their great stability. The range of the strong force extends only as far as the nucleus (10^{-15}m).

Electromagnetic force – This is responsible for the forces between charged particles such as protons and electrons, so can be attractive or repulsive, it is also responsible for electromagnetic emission and absorption. It also controls atomic structure, which allows atoms to bind together to make molecules.

The weak force – Acting at a shorter range than even the strong force, this interaction becomes particularly important during nuclear reactions, where it can switch the flavour of a quark, thereby changing a nuclear particle from one type to another. One example of this is beta decay, in which a neutrally charged neutron is converted into a positively charged proton, and a high energy (and negatively charged) electron is emitted. Another is a star's nuclear fusion reactions, which see the production of deuterium and then helium from hydrogen.

Gravitational force – This is the force we're most acutely aware of in our daily lives. It is, of course, the cause of stuff falling to the ground, and also of the Moon orbiting the Earth. It is only a noticeable force at large masses

– at the atomic level it's negligible due to its strength being 10^{40} times weaker than the strong force. Its range, however, is infinite.

As promised earlier, we're now ready to tackle what gauge bosons are. In the standard model, these are force carrier particles. They are responsible for three of the four fundamental forces (the standard model's main failing is that it can't actually explain gravity). Gluons carry the strong force; photons carry the electromagnetic force and W and Z bosons lie behind the weak force.

The Higgs boson, the final piece in the Standard Model jigsaw, was proposed the same year as the quark model, in 1964. It, along with its connected Higgs field, has played a key role in the development of Standard Model by explaining how each of the elementary particles come to have mass and also why photons and gluons don't. For a long time, it was the only particle in the model yet to be observed. However, on 4 July 2012, two separate teams of researchers at the Large Hadron Collider (LHC) at CERN presented experimental results which showed that after some rather committed searching they had almost certainly found evidence of a Higgs boson-like particle. The next question is whether it is the Higgs boson the standard model predicted or not. It's an exciting time to be a particle physicist.

> Look for knowledge not in
> books but in things themselves.
>
> William Gilbert, *De Magnete*

THE LARGE HADRON COLLIDER RAP

In the build-up to the official switch-on of the LHC at CERN, some scientists working there took it upon themselves to write, record and perform a rap in order to show the world what their research was all about. Within a week of it being posted on Youtube over two million people had viewed it and at the time of writing the figure stands at seven million. With lyrics by Kate McAlpine and music by Dr Will Barras, and featuring a variety of anonymous dancing scientists in numerous locations at CERN, the song is a brilliant piece of science communication (search for LHC rap on Youtube to watch the video). With special thanks to Kate McApline (www.katemcalpine.com) for allowing me to reproduce her lyrics:

Twenty-seven kilometers of tunnel under ground
Designed with mind to send protons around
A circle that crosses through Switzerland and France
Sixty nations contribute to scientific advance
Two beams of protons swing round, through the ring
*   they ride*
'Til in the hearts of the detectors, they're made
*   to collide*
And all that energy packed in such a tiny bit of room
Becomes mass, particles created from the vacuum
And then…

Chorus:

LHCb sees where the antimatter's gone
ALICE looks at collisions of lead ions
CMS and ATLAS are two of a kind
They're looking for whatever new particles they can
  find.
The LHC accelerates the protons and the lead
And the things that it discovers will rock you in the
  head.

We see asteroids and planets, stars galore
We know a black hole resides at each galaxy's core
But even all that matter cannot explain
What holds all these stars together – something else
  remains
This dark matter interacts only through gravity
And how do you catch a particle there's no way to see
Take it back to the conservation of energy
And the particles appear, clear as can be
You see particles flying, in jets they spray
But you notice there ain't nothin', goin' the other way
You say, 'My law has just been violated – it don't make
  sense!
There's gotta be another particle to make this balance.'
And it might be dark matter, and for first
Time we catch a glimpse of what must fill most of the
  known 'verse.
Because …

[Chorus]

Antimatter is sort of like matter's evil twin
Because except for charge and handedness of spin
They're the same for a particle and its anti-self
But you can't store an antiparticle on any shelf
Cuz when it meets its normal twin, they both
    annihilate
Matter turns to energy and then it dissipates
When matter is created from energy
Which is exactly what they'll do in the LHC
You get matter and antimatter in equal parts
And they try to take that back to when the universe
    starts
The Big Bang – back when the matter all exploded
But the amount of antimatter was somehow eroded
Because when we look around we see that matter
    abounds
But antimatter's nowhere to be found.
That's why…

[Chorus]

The Higgs Boson – that's the one that everybody talks
    about
And it's the one sure thing that this machine will sort
    out
If the Higgs exists, they ought to see it right away
And if it doesn't, then the scientists will finally say
'There is no Higgs! We need new physics to account for
    why
Things have mass. Something in our Standard Model
    went awry.'

But the Higgs – I still haven't said just what it does
They suppose that particles have mass because
There is this Higgs field that extends through all space
And some particles slow down while other particles race
Straight through like the photon – it has no mass
But something heavy like the top quark, it's draggin' its
    ***

And the Higgs is a boson that carries a force
And makes particles take orders from the field that is
    its source.
They'll detect it …

[Chorus]

Now some of you may think that gravity is strong
Cuz when you fall off your bicycle it don't take long
Until you hit the earth, and you say, 'Dang, that hurt!'
But if you think that force is powerful, you're wrong.
You see, gravity – it's weaker than Weak
And the reason why is something many scientists seek
They think about dimensions – we just live in three
But maybe there are some others that are too small to see
It's into these dimensions that gravity extends
Which makes it seem weaker, here on our end.
And these dimensions are 'rolled up' – curled so tight
That they don't affect you in your day-to-day life
But if you were as tiny as a graviton
You could enter these dimensions and go wandering on
And they'd find you …

[Chorus]

THE RELATIVE HARDNESS OF BEING

Devised by the German geologist Friedrich Mohs (1773–1839), the Mohs scale of mineral hardness arranges ten minerals in an order of increasing hardness, the difference being that a mineral on the list can be scratched by any below it.

Mohs hardness	Mineral
1 (softest)	Talc
2	Gypsum
3	Calcite
4	Fluorite
5	Apatite
6	Feldspar
7	Quartz
8	Topaz
9	Corundum
10 (hardest)	Diamond

The scale does not simply apply to these ten minerals – it can be used to assess the hardness of any solid. So, if a solid can be scratched by apatite but not fluorite then its Mohs hardness is between 4 and 5. Graphite (found in pencil lead) has a Mohs hardness of between 1 and 2. Anything up to 2.5 can typically be scratched by a fingernail, up to 4 by a coin and up to 6 by a knife. The scale is a relative one – the difference in absolute hardness between each mineral isn't consistent.

YOUR READING LIST FOR THIS WEEK IS ...

In 1998, the US publishing imprint Modern Library created a list of the 100 best non-fiction titles of the 20th century. It did not go uncriticised, particularly with regard to the ranking of the books (many of the judges weren't aware this was happening when they made their suggestions) but included in the list were a number of science and mathematics titles, if you're looking for a good read (once you're finished with this book, obviously). The number in brackets at the end of each entry refers to the book's ranking in the complete list of 100.

Silent Spring (1962), Rachel Carson (5)

The Double Helix (1968), James D. Watson (7)

The Lives of a Cell (1974), Lewis Thomas (11)

Principia Mathematica (1910, 1912, 1913), Alfred North Whitehead and Bertrand Russell (23)

The Mismeasure of Man (1981), Stephen Jay Gould (24)

The Art of the Soluble (1967), Peter B. Medawar (26)

The Ants (1990), Bert Hoelldobler and Edward O. Wilson (27)

On Growth and Form (1917), D'Arcy Thompson (34)

Ideas and Opinions (1954), Albert Einstein (35)

The Making of the Atomic Bomb (1987), Richard Rhodes (37)

Science and Civilization in China (1954), Joseph Needham (40)

The Structure of Scientific Revolutions (1962), Thomas S. Kuhn (69)

> *There exist only two kinds of modern mathematics books: ones which you cannot read beyond the first page and ones which you cannot read beyond the first sentence.*
>
> C.N. Yang, Nobel Prize winner

WHAT ARE YOU MADE OF?

This is the elemental composition of the average adult human body by mass (before any multivitamin and mineral tablets have been swallowed):

Oxygen 61%, Carbon 23%, Hydrogen 10%, Calcium 2.6%, Phosphorus 1.1%, Sulfur 0.2%, Potassium 0.2%, Sodium 0.14%, Chlorine 0.12%, Magnesium 0.027%, Silicon 0.026%, Iron 0.006%, Fluorine 0.0037%, Zinc 0.0033%, Rubidium 0.00046%, Strontium 0.00046%, Bromine 0.00029%, Lead 0.00017%, Copper 0.0001%, Aluminium 0.00009%, Cadmium 0.00007%, Boron 0.00007%, Barium 0.000003%, Tin 0.000002%, Manganese 0.000002%, Iodine 0.000002%, Nickel 0.000001%, Gold 0.000001%, Molybdenum 0.000001%,

Chromium 0.0000003%, Caesium 0.0000002%, Cobalt 0.0000002%, Uranium 0.0000001%.

WHY IS THE SKY DARK AT NIGHT?

Does this sound like a strange question? Well, it shouldn't. Named after a German astronomer (even though Kepler was actually the first to raise it) Olber's Paradox asks why the night sky isn't uniformly bright given, as the writer Edgar Allen Poe wonderfully put it in his prose poem *Eureka*:

> *Were the succession of stars endless, then the background of the sky would present us an uniform luminosity, like that displayed by the Galaxy – since there could be absolutely no point, in all that background, at which would not exist a star.*

In fact, the suggested explanation that Poe went on to give for this was very close to the one given over 50 years later by Lord Kelvin, which is still generally accepted. As Poe says:

> *The only mode, therefore, in which we could comprehend the voids which our telescopes find in innumerable directions, would be by supposing the distance of the invisible background so immense that no ray from it has yet been able to reach us at all.*

In other words, light from a great majority of stars in the universe hasn't yet reached us.

UNWEAVING THE RAINBOW

Many poets have railed against what they perceive to be the reductive nature of science. John Keats certainly didn't pull any punches when, in his poem *Lamia*, he referred to science as a 'cold philosophy' which would 'unweave a rainbow', thereby stripping the world of its beauty. But perhaps the most eloquent expression of this ungenerous view of science can be found in Walt Whitman's deceptively simple 'When I Heard the Learn'd Astronomer':

> *When I heard the learn'd astronomer,*
> *When the proofs, the figures, were ranged in columns*
>   *before me,*
> *When I was shown the charts and diagrams, to add,*
>   *divide, and measure them,*
> *When I sitting heard the astronomer where he lectured*
>   *with much applause in the lecture-room,*
> *How soon unaccountable I became tired and sick,*
> *Till rising and gliding out I wander'd off by myself,*
> *In the mystical moist night-air, and from time to time,*
> *Look'd up in perfect silence at the stars.*

The American physicist and bongo player Richard Feynman produced the perfect riposte to this view when he wrote in 'The Relation of Physics to other Sciences':

> *Poets say science takes away from the beauty of the stars – mere globs of gas atoms. Nothing is 'mere'. I too can see the stars on a desert night, and feel them. But do I see less or more? The vastness of the heavens*

stretches my imagination – stuck on this carousel my little eye can catch one-million-year-old light … It does not do harm to the mystery to know a little about it. For far more marvellous is the truth than any artists of the past imagined!

> *In science one tries to tell people, in such a way as to be understood by everyone, something that no one ever knew before. But in the case of poetry, it's the exact opposite!*
>
> **Paul Dirac**

THE NEW ATLANTIS

It might not come as a surprise to hear the roots of the Royal Society may lie in the work of Sir Francis Bacon. He is, after all, considered by many to be 'the father of experimental science'. But this belies the remarkable nature of this story.

According to the writer John Aubrey, William Harvey once derisively said of his patient Sir Francis Bacon that he wrote philosophy 'like a Lord Chancellor'. Bacon being, in fact, Lord Chancellor was part of the point. But it was also a criticism of a bureaucratic style of writing that Harvey felt Bacon had. While Harvey was clearly of the view that Bacon should have just concentrated on the day job, to Bacon his career was part of what provided him with the mental tools for the task he had in mind – to create an 'administration of learning'.

In *The New Atlantis*, a fictional utopian work published after his death, Bacon describes an institution called Salomon's House 'dedicated to the study of the works and creatures of God', the end of which is 'the knowledge of causes, and secret motions of things; and the enlarging of the bounds of human empire, to the effecting of all things possible'. The brethren of the house had access to a great deal of natural and artificial resources, enabling pretty much every area imaginable to be studied; there was a huge number of staff all contained within a very structured hierarchy, from 'servants and attendants' to 'novices and apprentices', to the particular specialised roles of the brethren themselves, including 'Merchants of Light', who travelled to other nations to retrieve their knowledge, and 'Pioneers' who devise and 'try new experiments'.

Bacon was essentially writing a manifesto for what he saw as the perfect scientific research establishment, the apotheosis of the systemisation of knowledge and learning. This was not a work for the general reader.

But it didn't need to be. The Prussian writer Samuel Hartlib, who spent most of his intellectual life in England, proved an influential advocate of Bacon's ideas. Hartlib believed passionately in educational reform and felt this could be best achieved via the real-life establishment of Salomon's House. He even went as far as identifying the Chelsea College as its possible location. He lobbied parliament to raise money for the project, but to no avail. However, among Hartlib's close circle of intellectual friends were the mathematician Sir William Petty and the natural philosopher Robert Boyle and the latter, at

THE SCIENCE MAGPIE

least, was a member with him of a 'philosophical college' that took 'the whole body of mankind for their care'. Both Petty and Boyle would be among the twelve people present at the inaugural meeting of the Royal Society in 1660, less than 40 years after Bacon's death.

The first few years of the society weren't easy – not everyone was convinced of its value and principles. Rather bizarrely for an organisation barely three years old, an official history by one of its fellows was commissioned to help present its credentials. Published in 1667, Thomas Sprat's *A History of the Royal Society* featured Francis Bacon in its frontispiece along with King Charles II and William Brouncker, the Society's first president. Through this, and the language used inside, it was clear how Baconian the Royal Society wanted to be seen as being. The implication that the utopian Salomon's House was now a reality was clear.

Approval wasn't unanimous and the Society still had to suffer its detractors. Jonathan Swift lampooned it superbly in his description of the grand academy of Lagado in his Gulliver's travels:

This academy is not an entire single building, but a continuation of several houses on both sides of a street, which growing waste, was purchased and applied to that use. I was received very kindly by the warden, and went for many days to the academy. Every room has in it one or more projectors; and I believe I could not be in fewer than five hundred rooms.

The first man I saw was of a meagre aspect, with sooty hands and face, his hair and beard long, ragged, and singed in several places. His clothes, shirt, and skin, were all of the same colour. He has been eight years upon a project for extracting sunbeams out of cucumbers, which were to be put in phials hermetically sealed, and let out to warm the air in raw inclement summers. He told me, he did not doubt, that, in eight years more, he should be able to supply the governor's gardens with sunshine, at a reasonable rate: but he complained that his stock was low, and entreated me 'to give him something as an encouragement to ingenuity, especially since this had been a very dear season for cucumbers'. I made him a small present, for my lord had furnished me with money on purpose, because he knew their practice of begging from all who go to see them.

During its 350 year history, the Royal Society has had some truly remarkable people as its presidents, many of whom have embodied the society's motto *'nullius in verba'* (roughly translated as 'take nobody's word for it'). These have included Christopher Wren, who was one of the twelve founding members, Samuel Pepys, Isaac Newton, Hans Sloane, Joseph Banks, Humphry Davy, T. H. Huxley, Joseph Lister, Ernest Rutherford and Howard Florey. Benjamin Franklin was the first American elected as a fellow of the society. However, it was, in many respects, still a mirror of society at large. It seems incredible today to learn that it wasn't until 1945, almost 300 years after its inception, that the first two female fellows of the Royal Society were elected – the crystallographer Kathleen Lonsdale and the biochemist Marjory Stephenson.

> *Benevolent maunderers stand up and say*
> *That black and white are but extremes of grey;*
> *Stir up the black creed with the white,*
> *The grey they make will be just right.*
>
> Written by T. H. Huxley, apparently in response
> to a church congress

A PLURALITY OF WORLDS IN EVERY EARRING

Margaret Cavendish (1623?–1673), Duchess of Newcastle upon Tyne, was formerly maid of honour to Queen Henrietta Maria, whom she accompanied into exile in Paris in 1644. There she met, among others, the philosopher René Descartes. An outspoken critic of women's subordination to men in society, and a flagrant rejecter of fashion, Cavendish was the first woman to visit the Royal Society, when in 1667, she attended scientific demonstrations by Robert Boyle and Robert Hooke. One can only wonder what Hooke thought about this – Cavendish had attacked his *Micrographia* in her *Observations upon Experimental Philosophy* (1666), famously commenting that 'the inspection of a bee, through a microscope, will bring him no more honey'. A committed writer, her best-known work is *The Blazing World*, an early example of science fiction, which describes a utopian society free of war and prejudice. A marble monument to Cavendish and her husband lies in Westminster Abbey and includes the line 'This Dutches was a wise wittie & learned Lady, which her many Bookes do well testifie'. In the following poem from the collection *Poems and Fancies* (1653), Cavendish employs common objects to explore the idea of there being a plurality of worlds, a theme she would visit more than once in her work.

Of Many Worlds in This World

Just like as in a Nest of Boxes round,
Degrees of Sizes in each Box are found:
So, in this World, may many others be
Thinner and less, and less still by degree:
Although they are not subject to our sense,
A World may be no bigger than Two-pence.
NATURE is curious, and such Works may shape,
Which our dull senses easily escape:
For Creatures, small as Atoms, may be there,

If every one a Creature's Figure bear.
If Atoms Four, a World can make, then see
What several Worlds might in an Ear-ring be:
For, Millions of those Atoms may be in
The Head of one small, little, single Pin.
And if thus small, then Ladies may well wear
A World of Worlds, as Pendents in each Ear.

Twenty years after the poem was published the Dutch scientist Antonie van Leeuwenhoek sent letters to the Royal Society describing his experiments with vastly improved microscopes he'd built. Over the following years he recorded many original observations including seeing what he termed 'animalcules' (bacteria) in plaque scraped from his teeth. An irony perhaps was that the inspiration for Leeuwenhoek's work had come from Hooke's *Micrographia*, which Cavendish had disparaged. Well, nobody's perfect.

Science was for Marx a historically dynamic, revolutionary force. However great the joy with which he welcomed a new discovery in some theoretical science whose practical application perhaps it was as yet quite impossible to envisage, he experienced quite another kind of joy when the discovery involved immediate revolutionary changes in industry and in historical development in general. For example, he followed closely the development of the discoveries made in the field of electricity and recently those of Marcel Deprez.

From Frederick Engels's eulogy at Karl Marx's Funeral

A LAW BY ANY OTHER NAME …

If I had to name three people synonymous with scientific laws I would likely respond Boyle, Hooke and Newton. The first of these, Boyle's law, is named after Robert Boyle (1627–1691), a founding member of the Royal Society, and concerns the relationship between the pressure and volume of a gas, as given by the equation:

$$pV = \text{constant}$$

However, on the continent the law is known by a different name, Marriott's law, after the French Abbot Edme Mariotte, who proposed his own version of the law. He was responsible for a number of other discoveries including the fact that the place where the optic nerve joins the eye's retina is a blind-spot, something you rarely realise because your brain fills in the missing spot, as well as using information from your other eye. It's not hard to find areas of disagreement between the French and the English, but in this instance, the English-speaking world is right – priority lies with Boyle who published his discovery more than ten years before the Abbot.

The Abbot also lost out somewhere else. 1967 saw the birth of the executive toy when actor Simon Prebble convinced Harrods to take his 'Newton's cradle'. In fact, it was Edme Mariotte who had first conducted experiments exploring the phenomena demonstrated by the cradle. In 1671 Mariotte presented his findings to the French Academy of Sciences and two years later published them. Newton acknowledged Mariotte's work fourteen years later in his *Principia*. Whether an executive toy named Mariotte's Cradle would have sold quite so well is open to conjecture.

> *Father of Chemistry and Uncle of the Earl of Cork*
> **Reported epitaph of Robert Boyle**

JUST NOT MY CUP OF TEA

In his book *Inflight Science*, Brian Clegg discusses whether it's possible to get a decent cup of tea while on an aeroplane. His answer is an emphatic no, for the simple reason that the cabin crew cannot boil water at a tea connoisseur's ideal of 100°C, because the air pressure in the cabin is lower than we generally experience in everyday life – the equivalent of roughly 6,000 feet above sea level. Since the boiling point of water (or any substance) drops in accordance with pressure, water heated in an aircraft cabin will reach boiling point before it gets to 100°C (it will boil at 93°C).

Tea isn't the only comestible which can have its quality affected by this feature of altitude, as Charles Darwin discovered in 1835 during his voyage on the Beagle:

> *Having crossed the [Piuquenes, Chile], we descended into a mountainous country … The elevation was probably not under 11,000 feet [3,350m] … At the place where we slept water necessarily boiled, from the diminished pressure of the atmosphere, at a lower temperature than it does in a less lofty country…. Hence the*

potatoes, after remaining for some hours in the boiling water, were nearly as hard as ever. The pot was left on the fire all night, and next morning it was boiled again, but yet the potatoes were not cooked.

Darwin's party would have found their water boiling at a temperature of approximately 88.7°C. This difference from the standard boiling point of 100°C clearly had a significant effect.

As a nation of tea drinkers, the British are lucky that their highest point of altitude is two-fifths that experienced by Darwin on the expedition. But so they're forewarned, the following table shows how the boiling point of water varies between various locations around the world.

	Altitude	Boiling point of water
Dead Sea	–423m	101.4°C
Sea-level	0m	100°C
Ben Nevis	1,344m	95.6°C
Cabin pressure of Boeing 767 at cruising altitude	2,100m	93.0°C
Mexico City	2,240m	92.6°C
Mont Blanc	4,810m	83.5°C
Kilimanjaro	5,895m	79.5°C
Everest	8,848m	68.0°C

The moral of the story is: if you really like your tea, don't fly on an aeroplane or try any impressive mountain climbing.

A PRIME DETERMINER

The sieve of Eratosthenes, named after the 3rd Century B.C. Greek mathematician, is a remarkably simple and effective way of sifting out the prime numbers (those only divisible by themselves and 1) in a sequence of natural numbers (positive integers). Here's how it works:

Cross out every second number after 2
Cross out every third number after 3
Cross out every fourth number after 4
Cross out every ... oh, you get the idea

This is what the 1 to 100 look like after doing this.

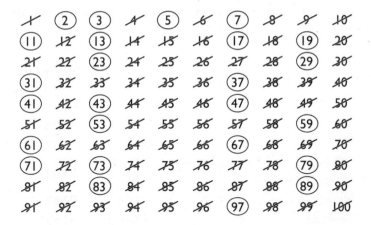

You may notice something rather interesting about the numbers remaining – they often come in pairs, differing only by 2, for example 3, 5 – 11, 13 – 17, 19 – 29, 31 and so on. Known as twin primes, calculations suggest that there are always such pairs, although so far no one has provided a proof of this.

The Prussian mathematician Christian Goldbach wrote a letter to Leonhard Euler in 1742 conjecturing that every even integer greater than 2 can be stated as the sum of two primes. For example $8 = 3+5$, $18 = 11+7$ and so on. No exception to the conjecture has so far been found but, as with twin primes, no proof has been given either.

Bloomsbury and Faber, the American and British publishers respectively of the novel *Uncle Petros and Goldbach's Conjecture*, by Apostolos Doxiadis, announced in 2000 that they would pay a prize of $1 million to anyone who proved the conjecture within two years of the book's publication. No one did and their respective bank managers were delighted.

FOR GOODNESS SAKE, SHOW YOUR WORKING

Pierre de Fermat (1601–1665) was a lawyer and amateur mathematician – amateur because he wasn't always interested in providing proofs for his theorems, as both of the examples of his work we are about to look at show. He spent most of his adult life in the French city of Toulouse and is perhaps most famous for his little and last theorems.

His little theorem, which can be expressed in a number of ways, is a method for finding out whether a number, p, is prime or not. If it is, and *a* is any positive integer, then

$$a^{p-1} - 1$$

will be divisible by p. For example, if we let p = 13 and $a = 2$, then

$$2^{12} - 1 = 4095$$

which is divisible by 13 (if 4095 is divided by 13, the answer we get is 315), so 13 is a prime number.

As was typical of Fermat, when he sent his theorem to a friend, he provided no proof because he feared it was too long. It would be another 100 years before Euler published one.

And it took over 350 years for a proof to be found to an assertion Fermat wrote in the margins of his copy of the Greek mathematician Diophantus' *Arithmetica,* and which became known as Fermat's last theorem ('To divide a cube into two other cubes, a fourth power or in general any power whatever into two powers of the same denomination above the second is impossible, and I have assuredly found an admirable proof of this, but the margin is too narrow to contain it'). In essence, it stated that no positive integers could satisfy the following equations:

$$x^n + y^n = z^n, \text{ where } n \geq 3$$

The special case of Fermat's last theorem is Pythagoras's theorem, when $n = 2$. Here, 3, 4 and 5 are possible solutions.

The problem perplexed mathematicians for centuries. Then, in June 1993, the English mathematician Andrew Wiles announced he had found a proof (in a document running to 200 pages). But within a few months an error was spotted. Wiles tried to fix it with the help of another

English mathematician, Richard Taylor. A year later, two papers were produced – a long one by Wiles and a short one by Wiles and Taylor. This time nobody disputed the proof, and the story was finally brought to a close. The prestigious journal *Annals of Mathematics* gave over its entire May 1995 issue to the proof.

> *I'm very well acquainted, too, with matters mathematical,*
> *I understand equations, both the simple and quadratical,*
> *About binomial theorem I'm teeming with a lot o' news,*
> *With many cheerful facts about the square of the hypotenuse.*
> *I'm very good at integral and differential calculus;*
> *I know the scientific names of beings animalculous:*
> *In short, in matters vegetable, animal, and mineral,*
> *I am the very model of a modern Major-General.*
>
> From 'I Am the Very Model of a Modern Major-General',
> *The Pirates of Penzance*, W.S. Gilbert

'THE DEGREE OF OPACITY'

The river Thames played a prominent role in the opening ceremony of the London 2012 Olympics – who can forget the iconic image of England's most capped outfield footballer David Beckham zooming up the river in a speed boat carrying the Olympic torch. Leaving aside anachronism, this would have been virtually impossible to do 150 years ago because of the overwhelming smell that Beckham would have experienced due to human contamination (I'm being polite here) of the river – it's hard to look dashing when you are suppressing the desire to retch – I know, I've tried. The summer of 1858 was particularly hot and the resulting smell so bad that it was labelled The Great Stink. No class of person was immune to its charms and it eventually led to a new sewage system devised by Joseph Bazalgette. However, that was still three years away when, in 1855, the eminent scientist Michael Faraday felt compelled to write to the editor of *The Times* regarding the state of the Thames. It was a very public statement from a man held in extremely high regard. His letter paints an unpleasantly vivid picture of the polluted state of the river at that time.

*

Sir, – I traversed this day by steamboat to the space between London and Hungerford bridges between half past one and two o'clock; it was low water, and I think the tide must have been near the turn. The appearance and the smell of the water forced themselves at once to my attention. The whole of the river was an opaque pale brown fluid. In order to test the degree of opacity, I tore up some white cards into pieces, moistened them so as to make them sink easily below the surface, and then dropped some of these pieces into the water at every pier the boat came to; before they had sunk an inch below the surface they were indistinguishable, though the sun shone brightly at the time; and when the pieces fell edgeways the lower part was hidden from sight before the upper part was underwater. This happened at St. Paul's-wharf, Blackfriars-bridge, Temple-wharf, Southwark-bridge, and Hungerford; and I have no doubt would have occurred further up and down the river. Near the bridges the feculence rolled up in clouds so dense that they were visible at the surface, even in water of this kind.

The smell was very bad and common to the whole of the water; it was the same as that which now comes up from the gully holes in the streets; the whole river was for the time a real sewer. Having just returned from out of the country air, I was, perhaps, more affected by it than others; but I do not think I could have gone on to Lambeth or Chelsea, and I was glad to enter the streets for an atmosphere which, except near the sinkholes, I have found much sweeter than that on the river. I have thought it a duty to record these facts that they may be brought to the attention of those who exercise power or have a responsibility in relation to the condition of our river; there is nothing figurative in the words I have employed, or any approach to exaggeration; they are the simple truth. If there be sufficient authority to remove a putrescent pond from the neighbourhood of a few simple dwellings, surely the river which flows for so many miles through London ought not to be allowed to become a fermenting sewer. The condition in which I saw the Thames may perhaps be considered as exceptional, but it ought to be an impossible state, instead of which I fear it is rapidly becoming the general condition. If we neglect this subject, we cannot expect to do so with impunity; nor ought we to be surprised if, ere many years are over, a hot season give us sad proof of the folly of our carelessness.

I am, Sir, your obedient servant,

Royal Institution, July 7.

M. FARADAY.

Two weeks after Faraday's letter was printed, *Punch* magazine published a witty caricature of 'Faraday giving his card to Father Thames' with their 'hope the Dirty Fellow [would] consult the learned Professor'.

REAL MOLECULES WITH SILLY NAMES

In 2008, Professor Paul May, a chemist from the University of Bristol, wrote *Molecules with Silly or Unusual Names*, an expansion of the website on the same subject he'd been maintaining since 1997. I strongly recommend you check it out, particularly if your sense of humour sometimes tends towards the puerile, as I'm afraid mine does. Included are the following gems:

Arsole

A ring molecule containing arsenic which came to prominence thanks to the *Journal of Organometalic Chemistry* paper, 'Studies in the Chemistry of Arsoles'. Does it help to say it was originally written in German?

Adamantane

Adam Ant was one of the most recognisable pop front men of the early 80s and co-writer of such classics as 'Prince Charming' and 'Stand and Deliver'. Unfortunately, the name of this chemical building block of diamond is unconnected, coming as it does from the Greek for indestructible, *adamas*.

Apollan-11-ol

This was first synthesised at the time of the Apollo 11 Moon landing. When drawn in two-dimensions, it looks like a child's drawing of a rocket, complete with fins and exhaust, and has an alcohol group attached to carbon 11 that provides the -11-ol suffix. It's certainly a more memorable name than the molecule it's based on: 1,4,4,7-tetramethyltricyclo[5.3.1.0$^{2.6}$]undecane.

Bastardane

Similar to adamantane, this molecule contains a structural deviation from standard hydrocarbon caged arrangements and so became known as bastardane, the unwanted child.

Diabolic acids

This molecule is a dicarboxylic acid with two carbon chains of different lengths in-between. The name comes from 'diabolos' derived from the Greek verb 'to mislead', because the acids prove particularly difficult to isolate.

Diurea

Perhaps you won't be surprised to hear this is used in the fertilizer industry.

Erotic acid

Not something to get you in the mood, this is really orotic acid. But Freud might have been right about his slips and possibly should have studied more chemists, as the acid has been misspelt so many times that the name erotic acid has now stuck.

Penguinone

So called because its two-dimensional structure resembles a penguin (sort of). It's a ketone, hence the –one ending.

Profilactin

The unrecognised name of the complex formed by the binding of the proteins profilin and actin, with the latter being relevant to muscle contraction, appropriately enough.

Uranate

This is the name of the uranium oxide anions such as UO_2^{2-}. Uranium nitrate is also known as uranyl nitrate. Maybe it's free during the day (the old ones are the best they say).

Windowpane

A hydrocarbon with the formula C_9H_{12}, the structure of which is four squares joined together to make a bigger square and hence resembling a window. It has never been synthesised but a version with one corner carbon missing, so that there were three squares joined together in an L-shape, has been, and was appropriately named 'broken window'.

> *A metallurgist is an expert who can look at a platinum blonde and tell whether she is virgin metal or a common ore.*
>
> **Anonymous**

THE MUSIC OF THE SPHERES

As we've seen elsewhere in this book, Johannes Kepler is a giant among giants in the history of science – an early believer in the Copernican system he would ultimately go beyond it to discover the true arrangement of the solar system. This was in part thanks to Danish astronomer Tycho Brahe, a brilliant astronomical observer whose own desire, ironically, was to prove a system in which the Earth most definitely didn't move.

Using Brahe's extremely accurate data, Kepler was able to determine that Mars travelled along an elliptical orbit, as opposed to the circle proposed by all systems before, and also that it moved with a regularly changing speed as it did so. A deeply religious man, this unexpected discovery posed a problem to Kepler – why would God produce a system based on ellipses when circles seemed so much more harmonious? His answer, it turned out, came in the form of the ancient Pythagorean belief in the 'music of the spheres'– the idea that a celestial soundscape is produced in the heavens by the motion of astronomical bodies.

Rather incredibly, by studying the speeds of each planet (there were six known at the time) at their aphelion (the point when a planet is farthest from the Sun and therefore when it is slowest) and perihelion (the point when a planet is closest to the Sun and fastest), Kepler worked out that the ratios of these pretty much matched those found in music, for example the ratio of the speed of Mars at perihelion and Earth at aphelion was the same as that between two tones in a perfect fifth. Remarkably, this work was of tangible benefit to science – a crucial consequence of Kepler's study of planet speeds was his third law. Here are all three laws:

1. A planet's orbit is elliptical with the Sun at one focus (an ellipsis has two foci).

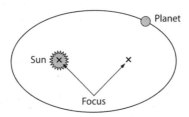

2. A line connecting a planet to the Sun sweeps out equal areas in equal periods of time.

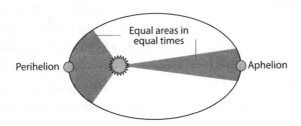

3. The square of a planet's period of revolution is directly proportional to the cube of its mean distance from the Sun. So, the larger the radius of its orbit, the greater its period of revolution, with an increase in the former effecting a greater proportional increase in the latter.

BRITAIN'S FIRST PROFESSIONAL FEMALE SCIENTIST

In 1762, William Herschel left his home town of Hanover to try to make his way as a musician. After four rather difficult years in London he moved to Bath, where he had some success, taking on pupils and performing in the famous Pump Rooms. Financially on a better footing than he'd ever been, he began to bring his siblings over to Bath, with his virtually uneducated sister Caroline arriving in 1772 to help run the household. William gave Caroline singing lessons and soon she was performing in public on a regular basis, taking the lead soprano role in works such as *Messiah*.

Ever since he'd arrived in England, William's interest in astronomy had been growing and, after his sister's arrival, it began to compete with music for his time. He built his own telescope, turned his basement into a foundry in an effort to produce better mirrors than he could otherwise source and published papers on his astronomical observations. Caroline started to help her brother more and more in his studies, spending hours polishing his telescopic mirrors, to the detriment of her musical career.

All this hard work paid off when William discovered the planet which would later become known as Uranus (William, in a bout of sycophancy, named it *Georgium Sidus*, the Star of George, after King George III, but the competing name of Uranus was suggested by the German astronomer Johann Elert Bode, and this eventually won out). His international fame was secured and the following year the king offered Herschel the position of royal astronomer at Windsor, with a pension of £200. He agreed and Caroline's musical career was effectively ended since she accompanied him, albeit in a more elevated position than mere household manager – William trained her to be an assistant astronomer.

Caroline was a meticulous observer and her speciality became nebulae and comets. On 1 August 1786 she found her first comet, and went on to discover seven more. In 1787 she was awarded her own pension from the King – £50. This made Caroline Britain's first professional female scientist. Her fame grew and in 1790 she received a letter addressed simply 'Mlle Caroline Herschel, Astronome Célèbre, Slough' from the director of the Paris Observatory. Her scrupulous work continued, even after her brother married and she was forced to take her own lodgings. In 1828 she was awarded a gold medal of the Astronomical Society of London. In later life she moved back to Hanover and wrote her memoirs. She died in 1848, aged 97.

The following is a poem written by the Swedish-American writer and artist Siv Cedering. It was first published in 1986 and won the Rhysling Award, founded by the Science Fiction Poetry Association. It is a wonderful modern interpretation of Caroline's life and a poignant reminder of some of the unsung heroines of science.

A Letter from Caroline Herschel (1750–1848)

William is away, and I am minding
the heavens. I have discovered
eight new comets and three nebulae
never before seen by man,
and I am preparing an index to
Flamsteed's observations, together with
a catalogue of 560 stars omitted from
the British Catalogue, plus a list of errata
in that publication. William says
I have a way with numbers, so I handle
all the necessary reductions and
calculations. I also plan
every night's observation
schedule, for he says my intuition
helps me turn the telescope to discover
star cluster after star cluster.

I have helped him polish the mirrors
and lenses of our new telescope. It is
the largest in existence. Can you imagine
the thrill of turning it to some new
corner of the heavens to see
something never before seen
from Earth? I actually like
that he is busy with the Royal Society

and his club, for when I finish my other work
I can spend all night sweeping
the heavens.

Sometimes when I am alone
in the dark, and the universe reveals
yet another secret, I say the names
of my long lost sisters, forgotten
in the books that record our science.

Aganice of Thessaly,
Hyptia,
Hildegard,
Catherina Hevelius,
Maria Agnesi

— as if the stars themselves could remember.

Did you know that Hildegard
proposed a heliocentric universe
300 years before Copernicus? That she
wrote of universal gravitation 500 years
before Newton? But who would listen
to her? She was just a nun, a woman.
What is our age, if that age was dark?
As for my name, it will also be
forgotten, but I am not accused
of being a sorceress, like Aganice,
and the Christians do not threaten to
drag me to church, to murder me, like they did
Hyptia of Alexandria, the eloquent young
woman who devised the instruments

used to accurately measure the position
and motion of heavenly bodies.

However long we live, life is short, so I
work. And however important man becomes,
he is nothing compared to the stars.
There are secrets, dear sister, and it is
for us to reveal them. Your name, like mine,
is a song.

Write soon,
Caroline

> ***Coeloum Perrupit Claustra***
> **[He broke through the barriers of the heavens]**
> **Epitaph to William Herschel, Church of**
> **St. Laurence, Upton-cum-Chalvey, Berkshire, UK**

THE BIRTH OF THE QUANTUM

In 1899, the American physicist Albert Michelson – who would win the Nobel Prize in Physics 1907 for his work in measuring the speed of light – made the sort of statement no scientist should ever be caught making. He declared 'the more important fundamental laws and facts of physical science have all been discovered, and these are now so firmly established that the possibility of their ever being supplanted in consequence of new discoveries is exceedingly remote'. This is the scientific equivalent of calling the *Titanic* 'unsinkable'.

Forty years earlier the German physicist Gustav Kirchoff had suggested the concept of the black body as a theoretical ideal designed to investigate the connection between the temperature of an object and the colour it radiated. For example, iron will always glow red at the same temperature and will change to orange, yellow and then white as it gets hotter, again at consistent temperatures. The black body provided the perfect way to measure this connection. Being black, it is able to absorb all electromagnetic radiation (all the different forms of light) that falls on it. The body, in turn, emits radiation which can then be measured. The hope was of establishing a formula that described how the intensity of radiation varied with wavelength at a given temperature (when iron glows red, the radiation emitted is most intense in the red part of the visible region of light).

One common problem for science is that it's often difficult to make ideal models like this into a reality. It wasn't until the 1890s that creating a black body became a reasonable possibility. In 1893, Willhelm Wien derived a formula that became known as 'Wien's displacement law' and which matched reasonably well with the experimental data being published at that time. The observed discrepancies – and there were some significant ones at the lower frequencies, the infrared end of the spectrum – were assumed by many to be the fault of the experiments rather than the law. The professor of theoretical physics at Berlin University, Max Planck, wanted to establish a proof of the law and set about working on this. While he did, the debate regarding differences between the law's predictions and actual data from experiments grew until significant cracks in the former began to show.

Using better quality data, Planck derived a formula that corresponded with it well and which he felt was an improvement of Wien's. The only problem was that it wasn't clear to anyone, especially Planck, what exactly the formula meant. There was still a lot more work to be done and Planck was forced to test nearly every scientific assumption he held.

After incorporating the ideas of Ludwig Boltzmann, who had produced a statistical description of the Second Law of Thermodynamics founded on probabilities many years earlier, Planck began to feel he was on the right track. But he was about to discover something totally unexpected. After toiling away he hit on a formula that appeared to match theory and data, which he first presented to the world on 14 December 1900. It was the surprisingly simple, but also earth-shattering:

$E = h\nu$

Where E is energy, ν is frequency and h is a constant. The stunning implication of this formula was that energy, like matter, was made up of very small, discrete units, which Planck called 'energy quanta'. So, energy increased and decreased in minute steps similar to walking up and down stairs. Up until that point, an increase and decrease in energy had been thought of as being smooth and continuous, like rolling up and down a perfect hill. h became known as Planck's constant and with a value of 6.626×10^{-34} J s^{-1} is one of the smallest quantities in physics.

The quantum (singular of quanta) had been

unleashed on the world. It was built on by Albert Einstein in 1905 in his paper on the photoelectric effect and extended by Niels Bohr in 1913 in his work on the structure of the atom. This work would eventually develop further into quantum mechanics and quantum electrodynamics. Without these quantum breakthroughs, and others, today we'd have no transistor, superconductor or laser to name but three examples. And I wouldn't be writing this book on a computer. Max Planck's Nobel Prize in Physics 1918 was justly deserved.

> *Scientists have odious manners,*
> *except when you prop up their theory;*
> *then you can borrow money off them.*
> Mark Twain

THE DEFINITION OF A PLANET

After the discovery of Pluto in 1930, our solar system was said to comprise nine planets, a fact that could be rather nattily remembered with the neat mnemonic *My Very Easy Method Just Speeds Up Naming Planets*.

However, a mini-crisis came about during the 1990s and 2000s when a number of other astronomical objects similar in size to Pluto – which was by then understood to be a member of a diffuse collection of objects called the Kuiper belt – were discovered. This then led to a fierce debate regarding what exactly constituted a planet.

To resolve the issue, the International Astronomical Union (IAU) produced the following definition:

'A planet is a celestial body that

(a) is in orbit around the Sun,

(b) has sufficient mass for its self-gravity to overcome rigid body forces so that it assumes a hydrostatic equilibrium (nearly round) shape, and

(c) has cleared the neighbourhood around its orbit.'

Pluto fails to satisfy the final condition and was downgraded as a result (it, and similar objects, are now known as plutoids). So, our solar system now contains only eight planets and a new mnemonic was born:

My Very Educated Mother Just Served Us Nachos

THE INTERSTELLAR PIONEER

Pioneer 10 was launched on 2 March 1972 and would become the first spacecraft to pass through the asteroid belt found between Mars and Jupiter, and the first to get up close to Jupiter. It was also the first to leave the Solar System and head into deep space. Because it was going further than anything man had made before, attached to it was a plaque created by the science writer and astrophysicist Carl Sagan, along with his wife Linda Salzman Sagan and the astrophysicist Frank Drake. Its purpose was to convey who, when and from where it had come to whoever might discover Pioneer 10 during its journey.

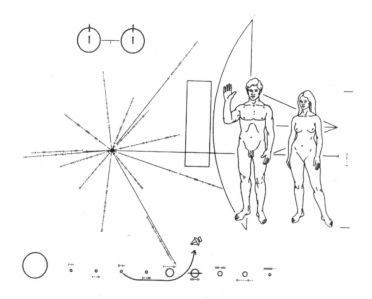

A week before the spacecraft's launch, Nasa described the plaque's symbology as follows:

The Pioneer [10] spacecraft, destined to be the first manmade object to escape from the Solar System into interstellar space, carries this pictorial plaque. It is designed to show scientifically educated inhabitants of some other star system, who might intercept it millions of years from now, when Pioneer was launched, from where, and by what kind of beings. (With the hope that they would not invade Earth.) The design is etched into a 6 inch by 9 inch gold-anodised aluminum plate, attached to the spacecraft's antenna support struts in a position to help shield it from erosion by interstellar dust. The radiating lines at left represent the positions of fourteen pulsars, a cosmic source of radio energy, arranged to indicate our sun as the home star of our civilisation. The '1-' symbols at the ends of the lines are binary numbers that represent the frequencies of these pulsars at the time of launch of Pioneer [10] relative to that of the hydrogen atom shown at the upper left with a '1' unity symbol. The hydrogen atom is thus used as a 'universal clock', and the regular decrease in the frequencies of the pulsars will enable another civilisation to determine the time that has elapsed since Pioneer [10] was launched. The hydrogen is also used as a 'universal yardstick' for sizing the human figures and outline of the spacecraft shown on the right. The hydrogen wavelength, about 8 inches, multiplied by the binary number representing '8' shown next to the woman gives her height, 64 inches. The figures represent the type of creature that created Pioneer. The man's hand is raised in a gesture of good will. Across the bottom are the planets, ranging outward from the Sun, with the spacecraft trajectory arching away from Earth, passing Mars, and swinging by Jupiter.

Contact was lost with Pioneer 10 in 2003 when it was some 12×10^6 km from Earth (80 AU). The ship is now ghosting its way in the general direction of Aldebaran, a star over 60 light years away in the constellation Taurus. It will take the spacecraft more than 2 million years to reach it. A second plaque of the same design was attached to Pioneer 11, launched just over a year after the first. This also journeyed to Jupiter before becoming the first spacecraft to explore Saturn and its rings up close. As with Pioneer 10, contact was lost as it headed into outer

THE SCIENCE MAGPIE

space where it is now moving towards Sagittarius and the centre of the galaxy.

THE TRUE MEASURE OF THINGS (PART 2)

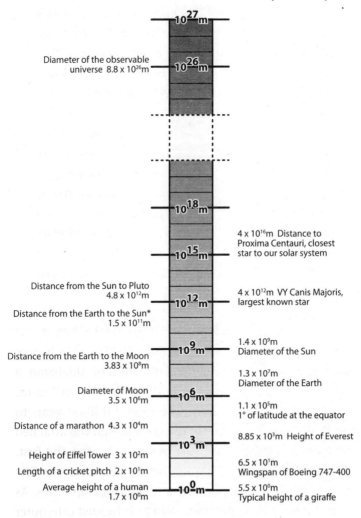

Diameter of the observable universe 8.8×10^{26}m

4×10^{16}m Distance to Proxima Centauri, closest star to our solar system

Distance from the Sun to Pluto 4.8×10^{12}m

4×10^{12}m VY Canis Majoris, largest known star

Distance from the Earth to the Sun* 1.5×10^{11}m

1.4×10^{9}m Diameter of the Sun

Distance from the Earth to the Moon 3.83×10^{8}m

1.3×10^{7}m Diameter of the Earth

Diameter of Moon 3.5×10^{6}m

1.1×10^{5}m $1°$ of latitude at the equator

Distance of a marathon 4.3×10^{4}m

8.85×10^{3}m Height of Everest

Height of Eiffel Tower 3×10^{2}m

Length of a cricket pitch 2×10^{1}m

6.5×10^{1}m Wingspan of Boeing 747-400

Average height of a human 1.7×10^{0}m

5.5×10^{0}m Typical height of a giraffe

NATURE *AND* NURTURE

For a number of years, we, the general public, were under the impression that we were effectively slaves to our genes and there wasn't much we could do about the hand nature had dealt us. Edward O. Wilson, father of sociobiology (the study of evolution's influence on social behaviour), famously paraphrased the Victorian writer Samuel Butler when he said 'People are DNA's way of making more DNA' (Butler had originally said 'a hen is only an egg's way of making another egg'). This gene-centred view has, over recent years, taken a bit of a battering, in part due to the rise of epigenetics, which examines how genes can be differently expressed depending on other factors.

Meaning 'upon the gene', epigenetics looks at the mechanisms involved in gene expression – in essence what switches them on and off and also what adjusts how 'loud' or 'quiet' they are. It's now known that what your DNA 'says' isn't enough – it very much matters how it's then read by your cells. It turns out that external factors, such as the environment you live in and the food you eat, can have a large impact on this reading of your genetic code through the placing of markers on your DNA (these don't affect its sequence though).

In *The Epigenetics Revolution*, Nessa Carey makes the analogy that DNA is like the script for a film or play. There have been literally thousands of performances of Shakespeare's Hamlet and, despite each using pretty much exactly the same script, each has interpreted what Shakespeare wrote in a slightly different way. Taking the analogy one step further, the director and actors may have made notes on the script, which facilitated their interpretation by possibly emphasising, downplaying or even cutting a particular passage. This, in a sense, is the influence nurture has on your DNA and how it's read.

To give an extreme example of what this means, when a queen honeybee founds a new colony, there will be literally thousands of eggs. When these hatch, the resulting bee larvae feed on royal jelly given to them by what are known as nurse bees. After three days this stops and the larvae are forced to make do with pollen and nectar, before eventually developing into worker bees, ready to start making honey. However, a very small number of larvae continue to be fed royal jelly and, astonishingly, these eventually grow into queen bees. Incredibly, there is nothing that genetically distinguishes queens from workers and yet two very physically different bees, with different lifespan and physiology, will have developed simply because of their diet. What food they ate during their growth affected the expression of the genes in their DNA. This appears analogous to temperature-dependent sex determination in certain reptiles, such as the crocodile. Here it's not possible to know what the sex of an offspring will be from its chromosomes. What affects the final outcome is the temperature of incubation of the egg at key moments in its development.

Something similar, but perhaps even more remarkable, that occurs in our own bodies is cell differentiation – that is, how stem cells know to become, say, specialised heart and kidney cells in their respective locations given they, like the bees above, start with the same genetic code.

In his 1957 book, *The Strategy of Genes*, the biologist Conrad Waddington proposed an epigenetic landscape to help visualise this process.

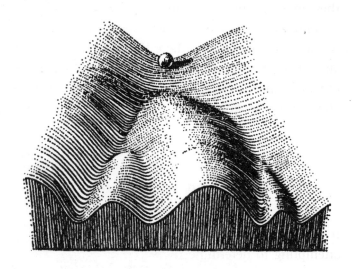

In his metaphor, he describes a ball about to start a journey down a hill. This represents a cell with the *potential* to specialise. As it progresses, it can roll down any one of a series of valleys and troughs until it eventually ends at the bottom in one of many possible final destinations – these valleys represent different courses of development, and the final destinations of the different possible cell types. By the time it reaches the bottom of the hill, the cell has specialised and become, for example, a heart, or lung, or kidney cell. To now get that cell to become a different type – to change a heart cell into a kidney cell, say

– would be extremely hard because the ball would have to be pushed over some rather steep inclines. The best option would be to push the ball back along its original journey in order to restore its potential, if this were possible. Scientists are now beginning to understand how they might be able to do this.

Epigenetics is one of the most important areas in science at the moment and has huge implications for us all. We are starting to get a much better understanding of what happens at the level of the gene and how this affects our lives. You're likely to hear a lot more about it in the future.

ONE CLEAR, UNCHANGED, AND UNIVERSAL LIGHT

It wasn't until the 17th century that people began attempting to determine the speed of light, c. Galileo describes an experiment to establish exactly that, in which lanterns a few kilometres apart are covered and then uncovered. This could never have worked due to the fact of c being such a large value, but the sentiment was an important one.

The Danish astronomer Ole Roemer made the first significant measurement in 1676. By observing successive eclipses of the moons of Jupiter by Jupiter itself, he came to a value of 214,000km s^{-1}. This is still quite some way off, but is at least in the right ball park, that is to say, it's very, very fast. The next significant step in determining the true value was made by the Englishman James

Bradley, who would go on to succeed Edmund Halley (after whom Halley's Comet is named) as Astronomer Royal. While attempting to detect stellar parallax, he came up with the remarkably accurate 301,000km s⁻¹. (Stellar parallax refers to the fact that stars are so far away from the Earth that you still have to look in the same direction to see them whether it's January or July, that is, at opposite sides of the Earth's circuit round the Sun.)

Leon Foucault, Albert Michelson and others made further advances during the late 19th century, and the 20th century saw various refinements until the current value of 299,792.458km s⁻¹ in a vacuum was adopted in 1983.

In 1905, Albert Einstein published his postulation that the speed of light was invariant; that it was the same whatever the motion of the observer. This is a key part of his special theory of relativity, which also includes the equation $E = mc^2$.

In September 2011, a team of scientists in Italy working on a project in collaboration with CERN announced results that appeared to show neutrinos travelling faster than the speed of light. If true, this would have turned physics on its head and heralded the end of Einstein's theories of relativity. In November the experiment was repeated and gave the same result. However, in March 2012 results from a different group at the same Italian laboratory showed neutrinos travelling at exactly the speed of light and it soon became clear there was a hardware problem with the first group's experiments and a huge sigh of relief could be breathed by all.

> *Twinkle, twinkle little star,*
> *I don't wonder what you are,*
> *For by spectroscopic ken*
> *I know that you are hydrogen.*
>
> Anonymous

ANOTHER HUE UNTO THE RAINBOW

We often hear of people suffering for their art but what about for science? Newton certainly did. During his investigations into colour he often experimented on his own eyes. This included staring at the Sun reflected in a mirror. It took him four days to recover his sight to an acceptable level, and he had recurring problems over the following months. Even more shockingly, Newton also physically interfered with his own sight. In one of his notebooks, he talks of inserting a bodkin (like a cross between an arrow and a needle) between his eye and its socket as near to the back of the eye as possible. He would then press so as to change the retina's curvature resulting in his seeing 'white darke & coloured circles' as he continued to vary the pressure and movement.

However, it was his rather more mundane investigations using prisms he'd bought at a local fair that showed that white light was made up of the colours of the rainbow. In 1800, over 100 years later, the British-German astronomer William Herschel began to investigate whether these colours corresponded to different temperatures.

Using part of a chandelier as his prism, Herschel split sunlight into its spectral colours. He then placed two 'control' thermometers on either side of the spectrum before using a third to measure the temperature of each colour. He soon discovered that as he went from violet to green to red, the temperature increased. This was an extremely interesting finding in itself but there was more to come. During the course of his experiments, the Sun had continued travelling on its course, which meant the split light had moved slightly as well. The thermometer had been measuring the red rays but was now just past them, yet Herschel was startled to find that the temperature measured was even higher – there appeared to be an invisible ray contained within sunlight. Via a well-constructed experiment, Herschel had accidentally discovered infrared radiation, a phenomenon that had been hypothesised in 1737 by Émilie du Châtelet. Like waiting for buses, the discovery of ultraviolet radiation quickly followed a year later.

James Clerk Maxwell's four famous equations subsequently explained that visible light, infrared and ultraviolet radiation were all electromagnetic waves travelling at the speed of light but with differing frequencies and wavelengths. The equations also predicted many other forms of electromagnetic radiation, all related by the equation:

$$\nu = \frac{c}{\lambda}$$

where ν is the frequency, c is the speed of light and λ is wavelength.

Over the course of the following 50 years, the discoveries of microwaves, radio waves, x-rays and gamma rays completed the picture, resulting in the following electromagnetic spectrum:

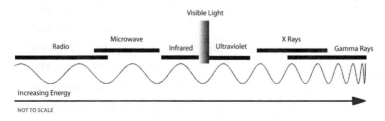

The advent of quantum theory, which now argued that energy could only come in chunks (this is effectively what a photon is – a packet of energy), produced a different version of the equation:

$$v = \frac{E}{h}$$

E refers to photon energy and h is Planck's constant. h has the incredibly small value of 6.626×10^{-34} and was something its originator Max Planck originally felt was an interim mathematical fudge to explain the results of an experiment, but which became a key part of this new science. This equation also tells us that the energy of an electromagnetic wave is related directly to its frequency. The higher the frequency, the higher the energy. Radio waves therefore possess the least energy whereas gamma rays have the highest. This is why gamma rays are so particularly harmful to human tissue, for example. (You may notice this equation doesn't tally with Herschel's findings regarding his thermometer's temperature increase as the frequency decreased – other factors were also involved).

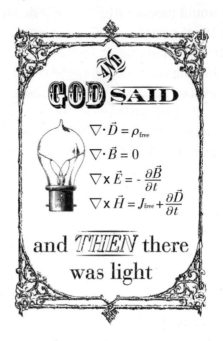

God said

$$\nabla \cdot \vec{D} = \rho_{\text{free}}$$
$$\nabla \cdot \vec{B} = 0$$
$$\nabla \times \vec{E} = -\frac{\partial \vec{B}}{\partial t}$$
$$\nabla \times \vec{H} = J_{\text{free}} + \frac{\partial \vec{D}}{\partial t}$$

and *THEN* there was light

ON A NEW KIND OF RAY

'If the hand be held between the discharge-tube and the screen, the darker shadow of the bones is seen within the slightly dark shadow-image of the hand itself … For brevity's sake I shall use the expression "rays"; and to distinguish them from others of this name I shall call them "X-rays"' – so wrote the German physicist Wilhelm Röntgen, winner of the first Nobel Prize in Physics in 1901, when he publicly announced the discovery he'd made on 8 November 1895.

Of huge scientific importance, not least due to the separate and equally significant transformation their discovery would produce within medicine and physics, Röntgen's 'rays' caused a public sensation right across Europe. By the beginning of February 1896 there was already a thriving trade in 'shadow photographs'. *Punch* magazine, ever one to chime with the public mood, published the following witty poem barely a month after Röntgen had submitted his preliminary communication to the Wurzburg Physico-Medical Society on December 28, 1895, and just two days after the journal *Nature* had printed a translation of it.

O, Röntgen, then the news is true,
And not a trick of idle rumour,
That bids us each beware of you,
And of your grim and graveyard humour.

We do not want, like Dr. Swift,
To take our flesh off and to pose in
Our bones, or show each little rift
And joint for you to poke your nose in.

We only crave to contemplate
Each other's usual full-dress photo;
Your worse than 'altogether' state
Of portraiture we bar in toto!

The fondest swain would scarcely prize
A picture of his lady's framework;
To gaze on this with yearning eyes
Would probably be voted tame work!

No, keep them for your epitaph,
These tombstone-souvenirs unpleasant;
Or go away and photograph
Mahatmas, spooks, and Mrs. B-s-nt!

CENTENARY ICONS OF SCIENCE AND TECHNOLOGY

To celebrate its centenary in 2009, the Science Museum in London selected ten 'centenary icons' of science and technology from right across its collection and invited people to vote for the one they felt was the greatest. The top place went to X-ray machines, which took 20% of the total vote. The following two positions were also taken by discoveries related to medicine.

1. X-ray machine
2. Penicillin
3. DNA double helix
4. Apollo 10 capsule
5. V2 Rocket Engine
6. Stephenson's Rocket
7. Pilot ACE Computer*
8. Steam Engine
9. Model T Ford
10. Electric telegraph

* The Pilot ACE (Automatic Computing Engine) was one of the first computers to be built in the UK and was precursor to the more ambitious ACE designed by Alan Turing.

MEASURING THE SPEED OF LIGHT USING CHOCOLATE, A MICROWAVE AND A RULER

As we have already seen in this book, it took over 300 years of experimentation and refinement to arrive at the figure for the speed of light which we use as standard today. That being the case, this method for determining that speed yourself might seem more than a little surprising. All you need is some kind of food that can melt (chocolate is good but you can also use marshmallows or cheese), a microwave oven, a microwave-safe dish to put the food in, and a ruler.

- Place the food on the dish.

- Remove the turntable from the microwave – it's important that the dish can't move.

- Put the dish in the microwave.

- Cook on a low heat until it's clear the food is beginning to melt in spots. Begin by trying 30 seconds. These spots relate to the peaks of the 'wave' – the distance between two peaks is half a wavelength.

- Once the melted spots appear, remove the dish and measure the distance between the centres of these spots. One distance should repeat again and again.

Now look on the microwave (it might be on the back) to find its frequency – this is typically 2.45GHz.

We know that $c = \lambda v$, or the speed of light = wavelength multiplied by frequency.

So, ν is the frequency of the microwave. If it's 2.45GHz then the figure you will use in your calculation will be 2,450,000,000 (whatever the frequency listed is, it will almost certainly be in gigahertz – 1GHz is 1,000,000,000 so make sure your calculation reflects that). You now need to multiply this by λ, which is double the distance you measured in metres (for example 15cm is 0.15 metres). See how close to the true speed of light, 299,792,458 ms^{-1}, you get.

'SPACE SHIFTED ABOUT LIKE A SWAN THAT CAN'T SETTLE'

On 15 June 1921, the English author D. H. Lawrence wrote to a friend to confirm he'd received the copy of Einstein's *Relativity: The Special and General Theory* he'd sent. The book was written with the non-specialist in mind, and Lawrence devoured its contents in a single day. A year later, he applauded the new theory, declaring himself to be 'very pleased with Mr. Einstein for knocking that eternal axis out of the universe. The universe isn't a spinning wheel. It is a cloud of bees flying and veering round'. Lawrence then shifted his attention from the universe to human relationships, proclaiming in his *Fantasia of the Unconscious* that 'we are in sad need of a theory of human relativity'.

Ironically, Lawrence would actually admit to not being able to grasp the new relativity, along with quantum theory, in his poem 'Relativity' included in his collection *Pansies* which was published in 1929, a year before his death.

Relativity

I like relativity and quantum theories
because I don't understand them
and they make me feel as if space shifted about like a
*   swan that can't settle,*
refusing to sit still and be measured;
and as if the atom were an impulsive thing
always changing its mind.

RELATIVELY SPEAKING

If we were ever compelled to answer the thorny question of who the greatest scientist who ever lived was, it would be extremely hard to look at anyone other than Isaac Newton and Albert Einstein as finalists.

1905 was Einstein's *annus mirabilis*. While still a patent clerk in Bern, Switzerland, he published five papers. One was on Brownian motion, and effectively put an end to the argument as to whether atoms truly existed or not; another was on the discovery of the photoelectric effect for which he won the Nobel Prize in Physics 1921; and perhaps most importantly, one was on his special theory of relativity, and included the line 'if a body gives off the energy L in the form of radiation, its mass diminishes by E/c^2'. This led to probably the most famous equation of all time, $E = mc^2$.

Eleven years later, Einstein expanded the special relativity paper so that it could take gravity and its effects into account, heralding the birth of the general theory of relativity. It was the apparent proof of this theory through the observation of a solar eclipse in 1919 that catapulted Einstein into the public consciousness. The following *New York Times* news report from the end of that year is full of charm and provides a rare glimpse into the early attempts to explain Einstein and his theory to the world at large.

*

EINSTEIN EXPOUNDS HIS NEW THEORY

It Discards Absolute Time and Space, Recognizing Them Only as Related to Moving Systems

IMPROVES ON NEWTON

Whose Approximations Hold for Most Motions, but Not Those of the Highest Velocity

INSPIRED AS NEWTON WAS

But by the Fall of Man from a Roof Instead of the Fall of an Apple

Special Cable to THE NEW YORK TIMES

BERLIN, Dec. 2. – Now that the Royal Society, at its meeting in London on Nov. 6, has put the stamp of its official authority on Dr. Albert Einstein's much-debated new 'theory of relativity', man's conception of the universe seems likely to undergo radical changes. Indeed, there are German savants who believe that since the promulgation of Newton's theory of gravitation no discovery of such importance has been made in the world of science.

When THE NEW YORK TIMES correspondent called at his

home to gather from his own lips an interpretation of what to laymen must appear the book with the seven seals, Dr. Einstein himself modestly put aside the suggestion that his theory might have the same revolutionary effect on the human mind as Newton's theses. The doctor lives on the top floor of a fashionable apartment house on one of the few elevated spots in Berlin – so to say, close to the stars which he studies, not with a telescope, but rather with the mental eye, and so far only as they come within the range of his mathematical formulae; for he is not an astronomer but a physicist.

It was from his lofty library, in which this conversation took place, that he observed years ago a man dropping from a neighboring roof – luckily on a pile of soft rubbish – and escaping almost without injury. This man told Dr. Einstein that in falling he experienced no sensation commonly considered as the effect of gravity, which, according to Newton's theory, would pull him down violently toward the earth. This incident, followed by further researches along the same line, started in his mind a complicated chain of thoughts leading finally, as he expressed it, 'not to a disavowal of Newton's theory of gravitation, but to a sublimation or supplement of it.'

When he read in the message from THE TIMES requesting the interview a reference to Dr. Einstein's statement to his publishers on the submission of his last book that not more than twelve persons in all the world could understand it, coupled with the editor's request that Dr. Einstein put his theory in terms comprehensible to a larger number than twelve, the doctor laughed good-naturedly, but still insisted on the difficulty of making himself understood by laymen.

'However,' he said, 'I am trying to talk as plainly as possible. To begin with the difference between my conception and Newton's law of gravitation: please imagine the Earth removed, and in its place suspended a box as big as a room or a whole house and inside a man naturally floating in the centre, there being no force whatever pulling him. Imagine, further, this box being, by a rope or other contrivance, suddenly jerked to one side, which is scientifically termed 'difform motion', as opposed to 'uniform motion'. The person would then naturally reach bottom on the opposite side. The result would consequently be the same as if he obeyed Newton's law of gravitation, while, in fact, there is no gravitation exerted whatever, which proves that difform motion will in every case produce the same effects as gravitation.

'I have applied this new idea to every kind of difform motion and have thus developed mathematical formulas which I am convinced give more precise results than those based on Newton's theory. Newton's formulas, however, are such close approximations that it was difficult to find by observation any obvious disagreement with experience.

'One such case, however, was presented by the motion of the planet Mercury, which for a long time baffled astronomers. This is now completely cleared up by my formulas, as the Astronomer Royal, Sir Frank Dyson, stated at the meeting of the Royal Society.

'Another case was the deflection of rays of light when

passing through the field of gravitation. No such deflections are explicable by Newton's theory of gravitation.

'According to my theory of difform motion, such deflections must take place when rays pass close to any gravitating mass, difform motion then coming into activity.

'The crucial test was supplied by the last total solar eclipse, when observations proved that the rays of fixed stars, having to pass close to the Sun to reach the Earth, were deflected the exact amount demanded by my formulas, confirming my idea that what so far has been regarded as the effect of gravitation is really the effect of difform motion. Elaborate apparatus and the closest and the most indefatigable attention to the difficult task enabled that English expedition, composed of the most talented scientists, to reach those conclusions.'

'Why is your idea termed "the theory of relativity?"' asked the correspondent.

'The term relativity refers to time and space,' Dr. Einstein replied. 'According to Galileo and Newton, time and space were absolute entities, and the moving systems of the universe were dependent on this absolute time and space. On this conception was built the science of mechanics. The resulting formulas sufficed for all motions of a slow nature; it was found, however, that they would not conform to the rapid motions apparent in electrodynamics.

'This led the Dutch professor, Lorenz, and myself to develop the theory of special relativity. Briefly, it discards absolute time and space and makes them in every instance relative to moving systems. By this theory all phenomena in electrodynamics, as well as mechanics, hitherto irreducible by the old formulae – and there are multitudes – were satisfactorily explained.

'Till now it was believed that time and space existed by themselves, even if there was nothing else – no Sun, no Earth, no stars – while now we know that time and space are not the vessel for the universe, but could not exist at all if there were no contents, namely, no Sun, Earth, and other celestial bodies.

'This special relativity, forming the first part of my theory, relates to all systems moving with uniform motion; that is, moving in a straight line with equal velocity.

'Gradually I was led to the idea, seeming a very paradox in science, that it might apply equally to all moving systems, even of difform motions, and thus I developed the conception of general relativity which forms the second part of my theory.

'It was during the development of the formulas for difform motions that the incident of the man falling from the roof gave me the idea that gravitation might be explained by difform motion.'

'If there is no absolute time or space, supposedly forming the vessel of the universe,' the correspondent asked, 'what becomes of the ether?'

'There is no ether, as hitherto conceived by science, which is proved by the well known experiment of the celebrated American savant, Michelson, showing that no influence by the motion of the Earth on the ether is perceptible through change in velocity of light, such as ought to be produced if the old conception were true.'

'Are you yourself absolutely convinced of the correctness of this revolutionary theory of relativity, or are there still any reservations?'

'Yes, I am,' Dr. Einstein answered. 'My theory is confirmed by the two crucial cases mentioned before. But there is still one test outstanding, namely the spectroscopic. According to my theory, the lines of the spectra of fixed stars must be slightly shifted through the influence of gravitation exerted by the very stars from which they emanate. So far, however, the results of the examinations have been contradictory; but I have no doubt of final confirmation even through this test*.'

Just then an old grandfather clock in the library chimed the mid-day hour, reminding Dr. Einstein of some appointment in another part of Berlin, and old-fashioned time and space enforced their wonted absolute tyranny over him who had spoken so contemptuously of their existence, this terminating the interview.

*Later experiments would show Einstein had been right about this also.

> *In science the credit goes to the man who convinces the world, not to the man to whom the idea first occurs.*
>
> **Sir Francis Darwin, botanist and son of Charles Darwin**

ROYAL SOCIETY PRIZE FOR SCIENCE BOOKS

First awarded in 1988, the Royal Society prize for science books was established 'with the aim of encouraging the writing, publishing and reading of good and accessible popular science books'. It has had a number of sponsors during that time and has been variously referred to as the 'Rhône-Poulenc Prize for Science Books' (1990–2000), the 'Aventis Prize for Science Books' (2001–2006), the 'Royal Society Prize for Science Books' (2007–2010) and the 'Royal Society Winton Prize for Science Books' (2011 to date). Except for the years 2009–2010, a companion junior book prize has run concurrently.

1988 – *Living with Risk*, British Medical Association

1989 – *Bones of Contention*, Robert Lewin

1990 – *The Emperor's New Mind*, Roger Penrose

1991 – *Wonderful Life*, Stephen Jay Gould

1992 – *The Rise and Fall of the Third Chimpanzee*, Jared Diamond

1993 – *The Making of Memory*, Steven Rose

1994 – *The Language of Genes*, Steve Jones

1995 – *The Consumer's Good Chemical Guide*, John Emsley

1996 – *Plague's Progress*, Arno Karlen

1997 – *The Wisdom of Bones*, Alan Walker and Pat Shipman

1998 – *Guns, Germs and Steel*, Jared Diamond

1999 – *The Man who Loved Only Numbers*, Paul Hoffman

2000 – *The Elegant Universe*, Brian Greene

2001 – *Mapping the Deep*, Robert Kunzig

2002 – *The Universe in a Nutshell*, Stephen Hawking

2003 – *Right Hand, Left Hand*, Chris McManus

2004 – *A Short History of Nearly Everything*, Bill Bryson

2005 – *Critical Mass*, Philip Ball

2006 – *Electric Universe*, David Bodanis

2007 – *Stumbling on Happiness*, Daniel Gilbert

2008 – *Six Degrees; Our Future on a Hotter Planet*, Mark Lynas

2009 – *The Age of Wonder*, Richard Holmes

2010 – *Life Ascending*, Nick Lane

2011 – *The Wavewatcher's Companion*, Gavin Pretor-Pinney

THE BOOK THAT CAUSED A PARADIGM SHIFT

1962 saw the publication of a history and philosophy of science monograph that would go on to sell well over 1 million copies and which would introduce to the world the concept of the paradigm shift. By looking at how science has been practised throughout history, *The Structure of Scientific Revolutions*, by Thomas S. Kuhn, seriously questioned the then standard view that science progresses by building incrementally on what has gone before and can therefore be said to be edging steadily towards a *true* description of the world.

Kuhn began by describing what he referred to as 'normal science'. An example being Newtonian mechanics, 'normal science' consists of a mature set of scientific ideas that subsequently determine what research the scientific community does, typically by looking for what they expect to find. However, over time, the number of unexplained experimental anomalies within this 'paradigm' grows and a crisis can occur. This, in turn, can lead to a revolution. In these circumstances, a collection of scientists typically continue to work within the current paradigm, attempting to rescue it. At the same time, a splinter group will probably help to give birth to an entirely new paradigm, which over time might gain the support of more and more scientists until, at some point, science undergoes a paradigm shift, whereby the worldview that was initially revolutionary becomes the generally accepted status quo. A classic example is Copernicus's proposal that the Earth revolved around the Sun. But this shift is rarely smooth. As Max Planck – who prompted one such revolution when his 'black body problem' resulted in the birth of quantum theory – put it, 'a new scientific truth does not triumph by convincing its opponents and making them see the light, but rather because its opponents eventually die, and a new generation grows up that is familiar with it'.

One consequence of Kuhn's book was that it seemed to suggest the notion of incommensurability in science – the view that concepts discussed in a new paradigm bore *no* connection with those in the previous one, that is, for example, that gravity in Einstein's general theory of relativity was *entirely* different to Newton's – they were talking about two different things (there was an argument that the mathematics underpinning them reinforced this view). If that was the case, then science couldn't exactly be said to be building on what had come before.

Instead, the inference was that it just moved to a completely different building altogether. Needless to say, this interpretation didn't go down terribly well within the scientific community.

Kuhn's ideas contained many problems of their own however, not least that they only really described the very big moments in the history of science – things like Copernicus, quantum theory and relativity, nearly all of which were centred around the physical sciences. According to the philosopher Ian Hacking, when the life sciences are taken into account (and their role within science has increased considerably since Kuhn wrote his book) Kuhn's diagnosis is much less convincing because these tend to be much less theory-driven, and much more analytical, than the physical sciences. But there can be little doubt of the impact *The Structure of Scientific Revolutions* has had – it provoked its own revolution in how the practice of science is understood.

> *First get your facts; and then distort them at your leisure.*
>
> **Mark Twain**

PRACTICAL POLITICS

In spring 1969, just months after Richard Nixon was elected president of the United States of America, a questionnaire was sent to over 60,000 American college and university professors that would provide researchers with more than 300 pieces of information on these individuals, including many relating to their political orientation. Among those who responded were 1,707 physicists, 1,884 chemists, 2,916 mathematicians, 812 geologists, 4,567 biological scientists, 2,395 faculty members in colleges of medicine and 4,382 engineers. Three years later, a paper detailing the general political differences between disciplines, and between high-achieving scientists and their more 'rank and file colleagues', was published in the journal *Science*. Some very interesting trends were observed.

Researchers discovered a clear correlation between particular disciplines and the political orientations of professors within them. So, social scientists and humanities professors were more liberal than scientists, who were more liberal than engineers. Within the sciences themselves, disciplines ranked as follows in terms of most liberal to most conservative:

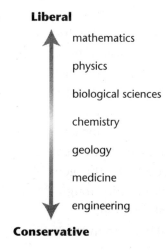

Liberal

mathematics

physics

biological sciences

chemistry

geology

medicine

engineering

Conservative

THE SCIENCE MAGPIE

The researchers felt their results showed that there was a relationship between how intellectual a discipline was and how liberal and change-orientated its practitioners were, that is, the more practically-based it was, the more conservative its practitioners. They also discovered that within disciplines, the higher achievers had more liberal political views; the more 'rank and file' more conservative views.

TYPES OF IONISING RADIATION

The type of radiation being referred to whenever radioactivity is mentioned is ionising radiation. It is, as the name rather heavily implies, radiation that can ionise, that is, produce electrically charged particles in anything it strikes, thereby changing the chemistry of that thing and making it more reactive. The effect of this on organisms can be significant.

We are exposed to ionising radiation every minute of every day, much of it in the form of background radiation. This includes cosmic rays, rocks in the ground, radon gas, water and food. Bananas, for example, contain naturally occurring potassium-40, a radioactive isotope of potassium. Incredibly, there is even something known as the 'banana-equivalent dose', which was created in an attempt to contextualise artificial radiation exposures, especially to the general public (an X-ray screening at a US airport is roughly two and a half times a banana-equivalent dose).

Once general background radiation is accounted for,

other sources include one's occupation (aeroplane crew, for example are exposed to a greater amount of cosmic rays), medical procedures (X-rays, CT scans etc.), industrial uses and the legacy of past nuclear disasters and explosions.

Types of ionising radiation include:

Alpha (α) particles – Identical to the *nucleus* of a helium atom, these particles contain two protons and two neutrons bound together. They are by far the heaviest of the types of radiation and are easily stopped by a sheet of paper or the first layer of the skin. However, they can be extremely dangerous to humans if a source is ingested. In late 2006, the Russian journalist and ex-KGB officer Alexander Litvinenko was poisoned with polonium-210, a radioactive isotope of polonium, possibly administered to him in a cup of tea. It decays by alpha particle emission only.

Beta (β) particles – These are high-speed electrons and, as such, are also high-energy. They can penetrate further than alpha particles – through about 1–2cm of skin – but can be stopped by a few millimetres of aluminium.

Gamma (γ) rays – These are a high-energy type of electromagnetic radiation and therefore have no mass. They are extremely penetrative and can pass through the human body causing great damage.

X-rays – Like gamma rays, although with a lower frequency, longer wavelength and less energy, these are a type of electromagnetic radiation and can pass through the human body. They can cause harm in large doses and so medical and dental use is tightly controlled.

Cosmic rays – In spite of their name, these are not actually rays but highly-charged subatomic particles, typically protons but also electrons and atomic nuclei. Their origins lie in a variety of sources located in outer space, such as exploding supernovae. It is estimated that about a million cosmic rays pass through our bodies each night while we sleep. Capable of penetrating far, particle physics experiments are performed deep underground so as to minimise any interference from the rays with the controlled particle collisions being performed.

> *If you're not part of the solution,*
> *you're part of the precipitate.*
> **Anonymous**

THE DOSE MAKES THE POISON

Our world has always been radioactive. The following pie chart, adapted from one produced by the World Nuclear Association – a body that 'represents the people and organisations of the global nuclear profession' – averages the sources of radiation we're exposed to every year. According to the WNA, 85% of the radiation we're exposed to comes from natural sources (though this will naturally depend on where you live and other factors).

Sources of radiation

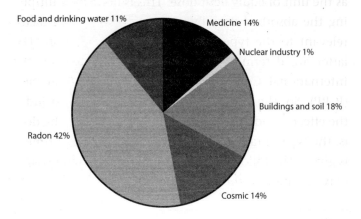

Food and drinking water 11%

Medicine 14%

Nuclear industry 1%

Buildings and soil 18%

Radon 42%

Cosmic 14%

There are a number of different radiation units. These include the SI (international system of units) derived becquerel (Bq), gray (Gy) and sievert (Sv).

The becquerel is a measure of the activity of a radio-active source. Here 1Bq equals 1 decay per second (1 decay equates to 1 emission of ionising radiation – for example, an α-particle).

The gray is the unit that measures the absorbed dose of radiation, and refers to the amount of energy absorbed as related to mass. 1 Gy equals 1 joule per kilogram of the mass of the thing doing the absorbing, such as the human body.

However, a measurement of an absorbed dose isn't necessarily the best way of assessing its biological effects, and therefore its danger. Different things, such as different human tissue or organs, are affected in different ways. As such, to best appreciate the risk, it's necessary to translate the absorbed dose into the sievert, known as the unit of equivalent dose. This is done by multiplying the absorbed dose by a radiation weighting factor relevant to the type and energy of the radiation. The latter are determined, and regularly updated, by the International Commission on Radiological Protection (ICRP). So, the sievert is used when we want to judge the effect of radiation on us. It matters how big the dose is, the type of radiation and over what period of time it is given. The following table compares different dosages (pay attention to the time periods involved).

Dose, millisieverts (mSv)	Event
10,000	Given as a single dose, this would prove fatal within weeks
6,000	Typical dosage recorded in those Chernobyl workers who died within a month
5,000	As a single dose, this would kill half of those exposed to it within a month
1,000	As a single dose, this would cause radiation sickness, including nausea and lower white blood cell count, though would not prove fatal
1,000	As an accumulated dosage this amount is estimated to cause a fatal cancer many years later in 5% of people
400	Maximum radiation levels recorded at Fukushima plant, per hour
350	Exposure of Chernobyl residents who were relocated after the blast in 1986

(continued)

Dose, millisieverts (mSv)	Event
250	Average annual background level at Ramsar in Iran (no identified health effects)
100	Lowest annual dose at which an increase in occurrences of cancer is clearly evident in recipients
20	Current annual limit for employees in nuclear industry
16	CT scan: heart
10	Full-body CT scan
9	Airline crew flying New York to Tokyo polar route, annual exposure
7	Average annual background radiation per person in Finland
2.2	Average annual background radiation per person in the UK
0.2	Chest X-ray
0.005	135g bag of Brazil nuts/dental X-ray

The general public has long been divided into two parts; those who think science can do anything, and those who are afraid it will.

Dixy Lee Lewis, marine biologist, former chairwoman of the Atomic Energy Commission and Washington state's first female governor

> *When it comes to atoms, language can be used only as in poetry.*
>
> Niels Bohr

TO DECAY WITH PRECISION

Uranium is probably the most famous of radioactive substances. Element number 92 in the periodic table and naturally occurring, it was through experiments on uranium that the phenomenon of radioactivity was first recognised in 1896 by French physicist Henri Becquerel. He shared the Nobel Prize in Physics 1903 with Marie and Pierre Curie. Uranium was also the radioactive substance involved in the first controlled nuclear chain reaction and lay at the heart of the atomic bomb dropped on Hiroshima. As such, it provides the perfect gateway into understanding some of the characteristics of radioactive materials.

Over 99% of all natural uranium comes in the form of its isotope uranium-238 (the isotopes of an element differ only in the number of neutrons in their nuclei – there are five other isotopes of uranium one of which, uranium-234, is listed in the decay chain below). Uranium-238 is unstable and gradually decays. It has a half-life (the time it takes for a mass of it to decrease by half) of more than 4 billion years. Furthermore, uranium decays naturally through a series of fourteen steps, with radioactive emission occurring at each one, finally resulting in lead-206. Despite not resulting in gold, this alchemical transmutation of one substance into another is one of nature's most fascinating processes.

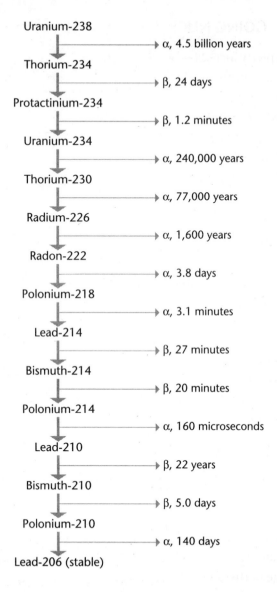

Uranium-238
→ α, 4.5 billion years
Thorium-234
→ β, 24 days
Protactinium-234
→ β, 1.2 minutes
Uranium-234
→ α, 240,000 years
Thorium-230
→ α, 77,000 years
Radium-226
→ α, 1,600 years
Radon-222
→ α, 3.8 days
Polonium-218
→ α, 3.1 minutes
Lead-214
→ β, 27 minutes
Bismuth-214
→ β, 20 minutes
Polonium-214
→ α, 160 microseconds
Lead-210
→ β, 22 years
Bismuth-210
→ β, 5.0 days
Polonium-210
→ α, 140 days
Lead-206 (stable)

α and β indicate alpha and beta decay. The times shown are half-lives.

GOING NUCLEAR IN A SPORTS HALL

It's a pretty incredible fact that the world's first nuclear reactor was constructed in a racquets court under the spectator stands of the University of Chicago's at that time recently closed football stadium, Stagg Field (today's stadium is several blocks northwest of where the original stood).

Built in the late autumn of 1942, Chicago Pile-1, as it was known, consisted of layers of solid graphite blocks with layers containing uranium metal and/or uranium oxide fuel in between them. Interspersed at regular intervals were horizontal shafts in which three sets of 'control rods', made of more graphite but also cadmium plated, could be removed or reinserted. It was these rods that allowed the team of scientists to manipulate the anticipated nuclear chain reaction. Leading the project was the Italian physicist Enrico Fermi, who had emigrated to the US four years earlier after picking up his Nobel Prize in Sweden.

By 2 December 'the pile' was ready to be tested. Beginning at 9.45am, over the course of the following six hours the control rods were slowly and carefully withdrawn by small degrees. At 3.36pm the reaction became self-sustaining and was allowed to continue for a further 28 minutes. Then Fermi instructed the main control rod to be reinserted and the reaction stopped. It was a remarkable achievement.

Later that day, Arthur Compton, the man in overall charge of the project, phoned James B. Conant, president of Harvard, informing him 'the Italian navigator ha[d] landed in the New World'. 'How were the natives?' asked

Conant. 'Very friendly,' was Compton's reply. But the joy in the achievement wasn't unanimous. Once most of those present had left for the day, one of Fermi's fellow scientists, Leó Szilárd, let him know he thought the day would go down as a 'black day in the history of mankind'.

AN UNHOLY TRINITY

Just over two and half years after Enrico Fermi led a team of scientists to produce the first controlled nuclear chain reaction, the first atomic bomb was detonated in the Jornada del Muerto desert, New Mexico, at 5.30am on 16 July 1945. Codenamed Trinity, it was the culmination of the gargantuan Manhattan Project, a scientific program costing over $2 billion dollars, the aim of which had been to develop the first nuclear weapon.

The project was based primarily at Los Alamos, some 230 miles to the north of Jornada del Muerto.

As *Trinity Site* by the National Atomic Museum makes clear, the public wasn't officially made aware of the test until the atomic bomb had been dropped on Hiroshima, Japan, on 6 August. But many people were already aware that something extremely significant had taken place. The light from Trinity's blast could be seen across the entire state of New Mexico and beyond. Minutes later, the ensuing mushroom cloud climbed to over 38,000 feet high. The power of the detonation had been equal to 20,000 tonnes of TNT, or the payload of about 2,000 of the US's largest Second World War bomber, the B-29 Superfortresses. The bomb dropped on Hiroshima was approximately four-fifths as powerful. Trinity destroyed all living things within a mile.

William L. Laurence, a reporter for the *New York Times* and the Manhattan Project's official journalist, described the test with a terrible beauty when he wrote:

Just at that instant there rose as if from the bowels of the earth a light not of this world, the light of many suns in one. It was a sunrise such as the world had never seen, a great green supersun climbing in a fraction of a second to a height of more than eight thousand feet, rising ever higher until it touched the clouds, lighting up earth and sky all around with dazzling luminosity. Up it went, a great wall of fire about a mile in diameter, changing colors as it kept shooting upward, from deep purple to orange, expanding, growing bigger, rising as it

*was expanding, an elemental force freed from its bonds
after being chained for billions of years.*

William L. Laurence, *New York Times*, September 26,
1945, Account of the Trinity Test on 16 July 1945

IN COLD BLOOD

Hypothermia is defined as a person having a core temperature below 35°C, and is extremely dangerous. The list of signs and symptoms below is meant only as a general guide as there is a lot of variation between individuals. For example consciousness can be lost after a drop of only 4°C in the core temperature and death can occur with a higher temperature than 25°C.

Core temperature °C	Signs and symptoms
36–37°C	
35–36°C	Goosebumps. Hands numb. breathing and pulse quickened
34–35°C	Movement slow. Mild confusion.
32–34°C	Shivering typically stops. Thinking slow and speech slurred.
30–32°C	Unable to walk. Pupils dilated
28–30°C	Semi-conscious. Slow pulse and irregular heartbeat.
25.5–28°C	Unconscious
20°C	Heart stops
17°C	No electrical activity in the brain
13°C	Lowest recorded temperature that a person (a child) has recovered from

MORE THAN THE GERM OF AN IDEA

Louis Pasteur (1822–1895), perhaps France's greatest ever scientist, was a driving force in the development of the germ theory of disease. His interest in microorganisms began during his work on fermentation, the study of which would eventually lead to the process bearing his name: pasteurisation. It was first used on wine, only later being adopted by milk producers.

Pasteur effectively ended the prevailing theory of spontaneous generation (in truth, his results were less conclusive than he presented them) before turning his attention to the mechanism behind contagious diseases, those afflicting silkworms in southern France in particular. At the same time, others were investigating the cause of anthrax, a disease deadly to both livestock and humans.

By the time Pasteur finished investigating silkworms in 1870, the idea that microbes were the cause of anthrax was hanging in the balance, with many scientists willing to dismiss the idea. The principle problem concerned how outbreaks could suddenly occur in healthy livestock that had no contact with infected animals if tiny organisms were responsible. It is at this point that the other outstanding name in the history of germ theory, the small-town doctor Robert Koch, enters the story.

After meticulously studying the disease's life cycle, in 1876 Koch had the brilliant insight that spores of the disease could exist in the general environment for a very long time and *then* activate once inside an animal.

He extended the scope of his studies and soon came to the conclusion that specific microbes caused specific diseases. This was another step altogether and the rest of the world needed convincing. A big boost to the argument was to come from the development of vaccines.

In the first half of 1881, Pasteur confidently announced he and his team had managed to produce a vaccine for anthrax and willingly accepted a scientific trial to publicly prove its efficacy. Taking place at his challenger's own farm, 50 healthy sheep were collected together and half were injected with the vaccine. The remaining 25 were left as they were. Two weeks later, all 50 were injected with the microbe supposedly behind the disease. Within three days all 25 'control' sheep were dead, leaving only those that had been vaccinated alive. Played out so publicly, it was an extraordinary triumph and catapulted Pasteur into the highest echelons of the scientific pantheon.

But despite this success, the moves towards definitively proving that specific microbes caused specific diseases were essentially happening on an *ad hoc* basis, an inadequate method for settling that germ theory was correct once and for all. A more general approach was necessary. Here we come back to Robert Koch. After his work on anthrax, he decided that what was needed was a formalised list of requirements, which, if met, would prove that a particular disease was caused by a particular microorganism. This list is now known as Koch's postulates. He stipulated that:

1. The microorganism must be shown to be present in every case of the disease.

2. It must be possible to isolate the microorganism and grow it in pure culture.

3. When inserted into a healthy organism, the relevant disease should be produced.

4. The microorganism can again be isolated and grown once more.

These postulates created the necessary framework for determining the cause of certain diseases, but there was still a difficulty in obtaining pure cultures. At the same time that Pasteur was testing his vaccine, Koch hit on agar jelly, a derivative of seaweed, as the answer to this problem after learning of its ability to cultivate almost any bacteria while reducing the risk of contamination because of it being solid. Now everything was in place to enable germ theory to truly change the world. Within ten years the bacillae behind tuberculosis, cholera, rabies and typhoid had been identified, saving millions of lives.

> *Physicists use the wave theory on Mondays, Wednesday and Fridays, and the particle theory on Tuesdays, Thursdays and Saturdays.*
>
> **Sir William Henry Bragg**

A SHORT HISTORY OF THE ATOM

The word atom comes from the ancient Greek for 'uncuttable'. The idea of their existence was first proposed by Leucippus of Miletus (fl. 435 BC) and Democritus (fl. 410 BC). They believed there were indivisible particles of matter, atoms, that moved around a void, and that it was the different patterns these were arranged in that produced change in the world, which we then perceived. As Andrew Gregory makes clear in *Eureka!*, this distinction between the reality of the atomic world and how it then appears and is perceived by us is remarkable for its foresight. This view was refined and expanded by later thinkers, and reached its apotheosis in Lucretius's poem *De rerum natura* (On the Nature of Things), the rediscovery of which in the 15th century is the subject of Stephen Greenblatt's Pulitzer prize-winning *The Swerve: How the Renaissance Began*.

However, it wasn't until the very beginning of the 19th century, when John Dalton discovered atomic weight, that the concept of the atom can be truly said to have had practical consequences. It was only from this point that there was an understanding of chemical distinctness – the idea that each element is composed of its own atoms that in turn give the element its characteristic weight.

In 1897, J. J. Thomson announced the discovery of the electron. Less than ten years later, he discovered a correlation between the number of electrons in an atom and its atomic weight (in fact it was with atomic number but this wasn't yet understood). This led him to propose the 'plum pudding' model of the atom where

THE SCIENCE MAGPIE

the electrons are dotted throughout an atom, the rest of which is a sort of positively charged cloud, resulting in an overall neutral charge.

Believe it or not, the actual existence of the atom, as opposed to the attitude that it was merely a useful concept, wasn't accepted until 1905. It was then that Einstein showed, using the relatively new mathematics of statistical mechanics, that atoms were responsible for Brownian motion – the random movement of particles suspended in a liquid named after Robert Brown, who noted the phenomenon over 70 years earlier when observing pollen particles in water.

The plum pudding model didn't last long though. In 1911, Thomson's former student Ernest Rutherford announced a new model of the atom after interpreting the startling results of an experiment performed two years earlier. In this experiment a very small number of alpha particles had bounced back when fired at gold foil, something that shouldn't have happened if the plum pudding model was correct – if an atom was a kind of cloud there should have been nothing for a particle to bounce off. As Rutherford famously remarked, 'it was almost as incredible as if you had fired a 15-inch shell at a piece of tissue paper and it came back and hit you'. In his new model the atom now contained a tiny positively charged nucleus which was responsible for nearly all the atom's mass – it was when an alpha particle hit this that it rebounded. Electrons 'orbited' the nucleus like planets round the Sun. This is the view of the atom that we still learn at school.

One problem with Rutherford's model was the question of electrons staying in orbit. According to classical mechanics, an electron should *gradually* lose energy resulting in it spiralling into the nucleus. Niels Bohr's answer in 1913 was to quantise the atom. He argued that there existed only very specific energy levels an electron could occupy. One way of imagining this is to think of a tennis ball, which represents an electron, and a series of steps it can sit on, which represent the energy levels. The higher up the steps, the higher the energy and vice versa. In the atom, for an electron to move between these levels, energy, in the form of electromagnetic radiation, needs to be emitted or absorbed. There could be no gradual loss of energy and the spiral problem was solved. It was for this work that Bohr was awarded the Nobel Prize nine years later.

For centuries a debate had raged as to whether light was a particle or a wave. By the beginning of the 20th century, it was generally accepted to be both. In 1924, not long after Bohr's award, the French physicist Louis de Broglie asserted in his PhD thesis that the electron was also guilty of this wave-particle duality. Erwin Schrödinger subsequently realised that every wave had a corresponding equation describing it, and began working on one to describe de Broglie's discovery. This was unleashed on to the world in 1926 in the form of the Schrödinger equation in which Ψ, the capital of the Greek letter psi, represents the wave function. Ψ was something he felt was intimately connected with the cloud-like distribution of an electron's electric charge. This was because Schrödinger's equation concerned the probability distribution of a particle – how likely a particular position was. The fact it was now not possible to treat an electron as something of which the movement could be accurately traced (because the concept of probability had entered the mix) led Werner Heisenberg to formulate his infamous uncertainty principle, in which it is impossible to know precisely both the position and the momentum (involving its velocity, mass and direction) of an electron.

> **Q:** Why are quantum physicists so poor at sex?
> **A:** Because when they find the position, they can't find the momentum, and when they have the momentum, they can't find the position.

WOULD THE REAL SCHRÖDINGER'S CAT PLEASE STAND UP

In 1935, the Austrian physicist Erwin Schrödinger published what has become arguably the world's most famous thought experiment. It involves a cat that is neither alive nor dead. If that sounds almost nonsensical, in a way it is meant to – the experiment is designed to draw attention to one of the strangest elements of quantum theory, and, as Niels Bohr once remarked that 'those who are not shocked when they first come across quantum theory cannot possibly have understood it'. At the 1927 Solvay conference an attempt was made to clarify the meaning of some recent studies in quantum theory. The view settled on became known as the Copenhagen Interpretation. At its heart lies the role of observation. In essence, Bohr and his followers' version of quantum theory inferred that sub-atomic events existed as probabilities that were resolved *only* when the event was observed in some way. So for example, a given particle might be in any one of a number of positions within a locale, there being a particular probability associated with each position. It is the actual act of looking to check where the particle is that then determines where it is. Let's just let the enormity of that sink in for a moment – the idea is that reality settles into a particular configuration *only when being looked at*.

For Schrödinger, the implied effect this sense of being in limbo at the sub-atomic level had on our everyday, macroscopic world was simply unacceptable. In response to correspondence with Einstein, who egged him on, Schrödinger's thought experiment was intended to highlight this absurdity. It goes like this:

> *A cat is penned up in a steel chamber, along with the following diabolical device (which must be secured against direct interference by the cat): in a Geiger counter there is a tiny bit of radioactive substance, so small, that perhaps in the course of one hour one of the atoms decays, but also, with equal probability, perhaps none; if it happens, the counter tube discharges and through a relay releases a hammer which shatters a small flask of hydrocyanic acid. If one has left this entire system to itself for an hour, one would say that the cat still lives if meanwhile no atom has decayed. The first atomic decay would have poisoned it. The wave function of the entire system would express this by having in it the living and the dead cat (pardon the expression) mixed or smeared out in equal parts.*

Bohr was unrepentant, refusing to accept the thought experiment's conclusion and insisting the cat would be either dead or alive before any observation was made. It's perhaps an irony, given its purpose was to ridicule Bohr's views, that Schrödinger's cat has had the impact it has in the world beyond science. Literature in particular has mined this paradox, examples of which include Ursula le Guin's irreverent short story 'Schrödinger's Cat' and Frederik Pohl's science fiction novel *The Coming of the Quantum Cats*. Perhaps of lower aesthetic merit, but no less entertaining, it has also inspired some rather tongue-in-cheek poetry such as the following by Marilyn T. Kocher:

The Tale of Schrödinger's Cat

Schrödinger called his cat and said,
'You can be both alive and dead,
For a linear combination of states
Postulates two simultaneous fates.'
Poor shocked pussy could not say,
'I shall inform the SPCA.
Your pet theory seems to me
An ultraviolent catastrophy.'

What then did this kitty do?
She looked at him and said 'μ'

> 'Why', said the Dodo, 'the best way
> to explain it is to do it.'
> Lewis Carroll, *Alice in Wonderland*

TO A MEASURABLE INFINITY AND BEYOND

How many atoms are there in the universe? Believe it or not, regardless of how huge the figure must be, it's possible to express the number using just four digits:

$$\mathfrak{I} = 9^{9^{9^{9}}}$$

To write out that number in full would require $10^{369693100}$ digits, which, frankly, isn't possible when you consider the estimated age of the universe is less than 10^{18} seconds. $\mathfrak{I}$ came courtesy of Carl Friedrich Gauss who wanted a 'measurable infinity', which is a little odd given that in mathematics infinity is a concept rather than a number, denoting something that is beyond any fixed bound and which can't be resolved by counting or measurement, even in theory.

The symbol for infinity, ∞, was introduced by the English mathematician John Wallis in 1655, although discussion of the concept of infinity can be traced back to the ancient Greeks.

TRUTH TO THEIR FICTIONS

Given the large number of people who work in science, it's perhaps surprising that more scientists don't appear in fiction. In part inspired by the website www.lablit.com, which is dedicated to promoting 'the culture of science in fiction and fact', here is a list of titles in which they do. For better or worse, the list excludes any title that might be considered science fiction, such as *Brave New*

World, in an attempt to ground the list as much as possible in the 'real world'. I appreciate I haven't necessarily wholly succeeded in achieving this.

Brazzaville Beach, William Boyd

In this book an ecologist travels to Africa to study primates and escape her marriage to a brilliant but flawed mathematician. It won the McVitie and James Tate Black Memorial prizes.

The Gold Bug Variations, Richard Powers

Named by *Time* magazine as the best novel of 1991, this novel of ideas, with two love stories at its heart, draws on a variety of disciplines including genetics, computer programming, Flemish art and music composition.

Einstein's Dreams, Alan Lightman

Sometimes compared to Italo Calvino's *Invisible Cities* and Jorge Luis Borges's *Labyrinths*, this series of connected fables reveals the dreams Albert Einstein might have had when finalising his special theory of relativity in 1905.

Measuring the World, Daniel Kehlmann

Originally published in German, and an international bestseller, this is a good-humoured fictional account of a meeting between two giants of 18th century intellectual thought, the naturalist Alexander von Humboldt and the mathematician Carl Friedrich Gauss.

The Thing of Darkness, Harry Thompson

Longlisted for the Booker Prize, this retelling of the famous voyage of HMS Beagle, and the following

30 years, focuses on the relationship between the ship's captain, Robert Fitzroy, and its scientific passenger, Charles Darwin

Ship Fever and Other Stories, **Andrea Barrett**
A National Book Award winner in 1996, this collection of short stories set in the 19th century takes its inspiration from the practice of science, humanising it in the process.

Remarkable Creatures, **Tracy Chevalier**
A novel that brings to life Mary Anning, an early 19th century collector and seller of fossils in Lyme Regis, whose discoveries helped shape the new science of geology but who, due to her sex, received little credit for her finds.

Solar, **Ian McEwan**
A satirical novel on climate change written by an author who draws on science for many of his novels. It won the Bollinger Everyman Wodehouse Prize for Comic Fiction.

Thinks, **David Lodge**
A clever campus novel that examines the puzzle that is consciousness through the eyes of a male cognitive scientist and a female novelist who embark upon an illicit affair.

Two on a Tower, **Thomas Hardy**
Covering the relationship between the wife of a country squire and a poor astronomer ten years her junior, Hardy

wanted to 'set the emotional history of two infinitesimal lives against the stupendous background of the stellar universe'.

The Housekeeper and the Professor, Yoko Ogawa

This short novel tells the moving story of a young mother employed to look after a mathematician who suffered a head injury years earlier. As a result he can remember everything before the accident but can't hold on to anything since for more than 80 minutes. Every day, the housekeeper has to start from scratch, but soon learns to speak to him in the language of maths.

The Calcutta Chromosome, Amitav Ghosh

An imaginative tale that switches between the late 19th century and Sir Ronald Ross's work on the transmission of malaria and the modern day where a Ross enthusiast is desperate to get to the truth – which turns out to be stranger than fiction – behind how the Nobel Prize-winning discovery was made.

The New Men, C. P. Snow

The sixth book in Snow's Strangers and Brothers series, this 1954 novel centres on a group of scientists striving to produce controlled nuclear fission during the Second World War and the government's management of them.

Arrowsmith, Sinclair Lewis

Winner of the Pulitzer Prize in 1926, which the author declined, this story relates the journey through life of

Martin Arrowsmith, from his humble beginnings in the Midwest to scientific success despite the death of his wife from a plague he eventually finds a cure for.

Cantor's Dilemma, Carl Djerassi
By a renowned professor of chemistry, this novel delves into the murky high-stakes politics of scientific research and explores the question of just how far someone would go to secure a Nobel Prize.

Kepler, John Banville
The second book in his Revolutions trilogy, here Banville recreates the life of the astronomer Johannes Kepler, casting light on the period he lived in and his quest to understand the universe.

The Mind-Body Problem, Rebecca Goldstein
In this book a beautiful philosophy graduate rebelling against her strict orthodox Jewish upbringing finds herself adrift in life, eventually marrying a brilliant Princeton mathematician through a desire for reflected status. However, their respective world-views clash, mainly due to their positions on the famous mind-body problem (the difficulty of understanding the connection between our physical bodies and our mental worlds).

The Indian Clerk, David Leavitt
Based on the relationship between the respected British mathematician G. H. Hardy and the autodidact mathematical genius Srinivasa Ramanujan who, while a clerk in Madras, was invited to Cambridge by Hardy, who had read samples of his brilliant theorems.

***The Experimental Heart*, Jennifer Rohn**

By the founder of www.lablit.com, this contemporary romantic thriller focuses on two laboratory-based scientists and the external pressure upon them to achieve results.

THE LIFE OF π

Leonhard Euler introduced the Greek letter π to denote the ratio of a circle's circumference to its diameter during the 18th century. The concept already had an illustrious history by that point. Many ancient civilisations had determined useful approximations of π, including the Babylonians ($3\frac{1}{8}$) and the Egyptians ($\frac{256}{81}$). The Bible suggests a ratio of 3 in Kings 7:23. The ancient Greeks then came up with a variety of ratios, the best being the 2nd century mathematician Hipparchus's value of $\frac{377}{120}$, correct to four decimal places. During the first millennium, various astronomers in China and India suggested values correct to between three and six decimal places, including Arybatha in 499 A.D. and Tsu Chúng-chih (430–501 A.D.). But it wasn't until 1424, when Jamshid Masud al-Kashi published *al-Risala al-Muhitiyya* that a significant advance in the calculation of π was made – correct to sixteen decimal places.

Europe began to make a more serious effort towards the end of the 16th century when the race to calculate π properly started and to anyone who has the time and the inclination there is now no real maximum to the number of decimal places possible thanks to the modern computer.

In case you're hungry for it, here's a large piece of π:

3.

```
14159 26535 89793 23846 26433 83279 50288 41971 69399 37510
58209 74944 59230 78164 06286 20899 86280 34825 34211 70679
82148 08651 32823 06647 09384 46095 50582 23172 53594 08128
48111 74502 84102 70193 85211 05559 64462 29489 54930 38196
44288 10975 66593 34461 28475 64823 37867 83165 27120 19091
45648 56692 34603 48610 45432 66482 13393 60726 02491 41273
72458 70066 06315 58817 48815 20920 96282 92540 91715 36436
78925 90360 01133 05305 48820 46652 13841 46951 94151 16094
33057 27036 57595 91953 09218 61173 81932 61179 31051 18548
07446 23799 62749 56735 18857 52724 89122 79381 83011 94912
98336 73362 44065 66430 86021 39494 63952 24737 19070 21798
60943 70277 05392 17176 29317 67523 84674 81846 76694 05132
00056 81271 45263 56082 77857 71342 75778 96091 73637 17872
14684 40901 22495 34301 46549 58537 10507 92279 68925 89235
42019 95611 21290 21960 86403 44181 59813 62977 47713 09960
51870 72113 49999 99837 29780 49951 05973 17328 16096 31859
50244 59455 34690 83026 42522 30825 33446 85035 26193 11881
71010 00313 78387 52886 58753 32083 81420 61717 76691 47303
59825 34904 28755 46873 11595 62863 88235 37875 93751 95778
18577 80532 17122 68066 13001 92787 66111 95909 21642 01989
38095 25720 10654 85863 27886 59361 53381 82796 82303 01952
03530 18529 68995 77362 25994 13891 24972 17752 83479 13151
55748 57242 45415 06959 50829 53311 68617 27855 88907 50983
81754 63746 49393 19255 06040 09277 01671 13900 98488 24012
85836 16035 63707 66010 47101 81942 95559 61989 46767 83744
94482 55379 77472 68471 04047 53464 62080 46684 25906 94912
93313 67702 89891 52104 75216 20569 66024 05803 81501 93511
25338 24300 35587 64024 74964 73263 91419 92726 04269 92279
67823 54781 63600 93417 21641 21992 45863 15030 28618 29745
55706 74983 85054 94588 58692 69956 90927 21079 75093 02955
32116 53449 87202 75596 02364 80665 49911 98818 34797 75356
63698 07426 54252 78625 51818 41757 46728 90977 77279 38000
81647 06001 61452 49192 17321 72147 72350 14144 19735 68548
16136 11573 52552 13347 57418 49468 43852 33239 07394 14333
45477 62416 86251 89835 69485 56209 92192 22184 27255 02542
56887 67179 04946 01653 46680 49886 27232 79178 60857 84383
82796 79766 81454 10095 38837 86360 95068 00642 25125 20511
73929 84896 08412 84886 26945 60424 19652 85022 21066 11863
06744 27862 20391 94945 04712 37137 86960 95636 43719 17287
46776 46575 73962 41389 08658 32645 99581 33904 78027 59009
94657 64078 95126 94683 98352 59570 98258 22620 52248 94077
26719 47826 84826 01476 99090 26401 36394 43745 53050 68203
49625 24517 49399 65143 14298 09190 65925 09372 21696 46151
57098 58387 41059 78859 59772 97549 89301 61753 92846 81382
68683 86894 27741 55991 85592 52459 53959 43104 99725 24680
```

A SLICE OF INDIANA π

For over 2,000 years, people were determined to crack the problem of squaring the circle, that is, constructing a square the area of which is *exactly* the same as that of a given circle, using a straight edge and compass. Hopes that it might be possible were quashed when, in 1882, Ferdinand von Lindemann showed that π was, in fact, a transcendental number, meaning it wasn't the root of any algebraic equation. But this didn't put off the American physician Dr. Edwin J. Goodwin who, in 1897, wrote a bill for the state of Indiana with the intention of establishing a 'new mathematical truth' – his procedure for squaring the circle.

On 18 January 1897, House Bill 246 was submitted to the Indiana General Assembly. Amazingly, Goodwin actually wanted to copyright his solution, and therefore make money from it, declaring in the opening of the bill that it was to be 'offered as a contribution to education to be used only by the State of Indiana free of cost by paying any royalties whatever on the same, provided it is accepted and adopted by the official action of the Legislature of 1897'.

There was just one major flaw in Goodwin's new truth. Tucked away in the second section of the Bill was the 'fourth important fact' necessary to the method, that 'the ratio of the diameter and circumference is as five-fourths to four' or, rather, 3.2. So, Goodwin was essentially trying to establish in State law that $\pi = 3.2$, rather than 3.1459 etc. and then to make money from this 'fact'!

Astonishingly, the bill was passed on 5 February by 67 votes to none! It was then transferred to the Indiana Senate where its first reading by the Committee on Temperance was reported back favourably and was put formally before the Senate on 12 February. Thankfully, common sense was given a serendipitous opportunity – Professor Clarence Abiathar Waldo, a mathematician at Purdue University, had been visiting the Statehouse and overheard the General Assembly debate. He 'coached' the senators about the Bill, which was given short shrift and postponed indefinitely. This rather suggests it is still knocking about somewhere.

> **Where do you bury dead mathematicians? Asymmetry**

MEET TOM TELESCOPE AND FRIENDS

The Newbery Medal is the most prestigious children's book prize in the US today. Begun in 1922, and the first of its type, it is awarded by the American Library Association 'to the author of the most distinguished contribution to American literature for children' and named after the 18th century British publisher and bookseller John Newbery, widely recognised as the first person to specialise in children's books.

In April 1761, Newbery published *The Newtonian system of philosophy, adapted to the capacities of young gentlemen and ladies*, a popular science book probably aimed at children aged between ten and twelve. In the book, 'six Lectures [are] read to the Lilliputian Society, by Tom Telescope' which have been 'collected and methodised for the Benefit of the Youth of these Kingdoms'. The lectures cover the full range of 18th century natural philosophy, from matter and gravity to the five senses of man and continually make reference to objects with which young people would have been acquainted, such as balls and tops. What follows is the beginning of the first lecture in which Newton's 1st law of motion is introduced very clearly and concisely:

When a body is in motion, as much force is required to make it rest, as was required, while it was at rest, to put it in motion. Thus, suppose a boy strikes a trap-ball with one hand, and another stands close by to catch it with one of his hands, it will require as much strength or force to stop that ball, or put it in a state of rest, as the other gave to put it in motion; allowing for the distance the two boys stand apart.

No body or part of matter can give itself either motion or rest: and therefore a body at rest will remain so for ever, unless it be put in motion by some external cause; and a body in motion will move for ever, unless some external cause stops it.

This seemed so absurd to Master Wilson, that he burst into a loud laugh. What, says he, shall any body tell me that my hoop or my top will run for ever, when I know by daily experience that they drop of themselves, without being touched by any body? At which our little

Philosopher was angry, and having commanded silence, Don't expose your ignorance, Tom Wilson, for the sake of a laugh, says he: if you intend to go through my Course of Philosophy, and to make yourself acquainted with the nature of things, you must prepare to hear what is more extraordinary than this. When you say that nothing touched the top or the hoop, you forget their friction or rubbing against the ground they run upon, and the resistance they meet with from the air in their course, which is very considerable though it has escaped your notice. Somewhat too might be said on the gravity and attraction between the top or the hoop, and the earth; but that you are not yet able to comprehend, and therefore we shall proceed in our Lecture.

Remind me not to get on the wrong side of Tom Telescope.

The book was a huge success, with numerous editions, and as well as being published in Ireland and America, was translated into Dutch, Swedish and Italian. Many historians now feel it highly likely that John Newbery himself wrote the book. It was almost certainly publications like this that prompted Charles Lamb – apparently not much of a fan of science – to write to Coleridge in 1802 that 'Science has succeeded to poetry, no less in the little walks of children than with men. Is there no possibility of averting this sore evil?' If he were still alive today, Lamb might feel that both science *and* poetry have been 'averted', certainly as far as children are concerned.

THE SINS OF NEWTON

Isaac Newton, widely regarded as one of the greatest scientists to have ever lived, was a deeply religious man. An anti-Trinitarian, a heretical position during the 17th century, Newton spent a huge amount of time and effort attempting to uncover secret knowledge he felt was to be found in the sacred texts of ancient cultures. This included devising rules for interpreting the symbols of biblical prophecy – he was particularly interested in the books of Daniel and Revelations. Based on time periods in the former, he calculated the end of the world, the apocalypse, would not happen before 2060 A.D. We've a few years left at least.

More mundanely, although certainly not for Newton himself, as a nineteen year-old he recorded in a notebook on Whitsun 1662 all 49 sins he could recall committing up to that point. It makes fascinating reading and helps turn a scientific giant into someone very human. It also provides an insight into Newton's religious fervour, something that never left him. The following is a selection from the list:

Using the word (God) openly

Making a feather while on Thy day

Denying that I made it.

Squirting water on Thy day

Making pies on Sunday night

Threatning my father and mother Smith to burne them and the house over them

Wishing death and hoping it to some

Having uncleane thoughts words and actions and dreamese

Stealing cherry cobs from Eduard Storer

Denying that I did so

Setting my heart on money, learning, pleasure more than Thee

Punching my sister

Robbing my mother's box of plums and sugar

Calling Dorothy Rose a jade

Peevishness with my mother

Idle discourse on Thy day and at other times

Not loving Thee for Thy goodness to us

Not desiring Thy ordinances

Fearing man above Thee

Using unlawful means to bring us out of distresses

Not craving a blessing from God on our honest endeavors

Missing chapel

Beating Arthur Storer

Peevishness at Master Clarks for a piece of bread and butter

Striving to cheat with a brass halfe crowne

Twisting a cord on Sunday morning

From Newton's monument, Westminster Abbey:
H. S. E. ISAACUS NEWTON Eques Auratus, / Qui,
animi vi prope divinâ, / Planetarum Motus, Figuras,
/ Cometarum semitas, Oceanique Aestus. Suâ Mathesi
facem praeferente / Primus demonstravit: / Radiorum
Lucis dissimilitudines, / Colorumque inde nascentium
proprietates, / Quas nemo antea vel suspicatus erat,
pervestigavit. / Naturae, Antiquitatis, S. Scripturae, /
Sedulus, sagax, fidus Interpres / Dei O. M. Majestatem
Philosophiâ asseruit, / Evangelij Simplicitatem Moribus
expressit. / Sibi gratulentur Mortales, / Tale tantumque
exstitisse / HUMANI GENERIS DECUS. / NAT. XXV DEC.
A.D. MDCXLII. OBIIT. XX. MAR. MDCCXXVI

This can be translated as follows:
Here is buried Isaac Newton, Knight, who by a strength
of mind almost divine, and mathematical principles
peculiarly his own, explored the course and figures of
the planets, the paths of comets, the tides of the sea,
the dissimilarities in rays of light, and, what no other
scholar has previously imagined, the properties of the
colours thus produced. Diligent, sagacious and faithful,
in his expositions of nature, antiquity and the holy
Scriptures, he vindicated by his philosophy the majesty
of God mighty and good, and expressed the simplicity
of the Gospel in his manners. Mortals rejoice that there
has existed such and so great an ornament of the human
race! He was born on 25th December, 1642, and died on
20th March 1726.

MENDEL'S LAW OF SEGREGATION

Gregor Johann Mendel (1822–1884) was an Austrian monk whose experimental work laid the foundations of genetics. In 1866 he published the results of thousands of experiments in breeding peas, which he'd carried out in the gardens of his monastery. Mendel discovered there was a pattern discernible in the inheritance of characteristics such as flower colour in pea plants. This led him to formulate his law of segregation.

It states that two units control the inheritance pattern of a particular characteristic in an organism. We now know these to be alleles, different forms of a gene. An organism will carry two forms of the same gene, one from each of its parents. When germ (sex) cells are formed they receive only one of this pair. So because they're made by the fusion of a germ cell from each parent, the first offspring generation (the first filial generation) contains an allele from each parent. So if one of the parent plants has pure white flowers and the other pure purple flowers, the first filial generation will carry one allele for purple flowers and one for white flowers.

With the breeding of successive generations, it became clear to Mendel that certain traits could be either dominant or recessive and that the dominant would always mask the recessive. This explained particular patterns of inheritance and why traits could skip generations. For example, if we look at the inheritance of the relevant colour allele when pure purple and pure white pea plants are crossed we notice the colour white disappears in the second generation, because purple is

THE SCIENCE MAGPIE

the dominant trait. But white comes back in the third generation because it is possible for a plant to inherit two white alleles at this stage:

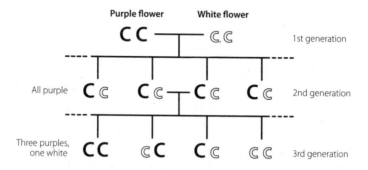

Charles Darwin may have provided the theory of evolution with the sound footing it needed but it was Gregor Mendel's work that enabled the study of the mechanisms that lay behind Darwin's theories.

THE MONKEY TRIAL

On 13 March 1925, the State of Tennessee passed an incredible law that stated evolution could not be taught in any university or public school they provided funding for. Anyone caught doing so faced a fine of up to $500, approximately half a teacher's annual salary. If anything, this felt like a challenge. The American Civil Liberties Union advertised for a teacher willing to make it known they'd contravened the law and taught the theory of evolution. John T. Scopes was the man who stepped forward. Thus began the Scopes Trial (also known as the Monkey Trial), one of the most famous court cases centring on science in the 20th century. It pitted the former US Secretary of State, William James Bryan, acting for the prosecution, against the acclaimed lawyer Clarence Darrow. It was a trial that would be transmitted across America by radio and, ironically, provided an excellent opportunity for the theory of evolution to be explained to millions of people.

On the eighth day of the trial, after the majority of the defence's arguments had been declared inadmissible by a clearly biased judge, the defence changed their plea to guilty and Scopes was fined $100. This gave the defence team the opportunity to take the case to the Tennessee Supreme Court where, two years later, the validity of the Butler Act was upheld but Scopes's sentence was reversed on a technicality – the minimum fine, which had been enforced, was double what could be imposed without a jury's assessment and the original 'trial judge [had] exceeded his jurisdiction in levying this fine'. The Act, which would not be repealed for another 40 years, is reproduced here:

The Butler Act

House Bill No. 185

AN ACT prohibiting the teaching of the Evolution Theory in all the Universities, Normals and all other public schools of Tennessee, which are supported in whole or in part by the public school funds of the State, and to provide penalties for the violations thereof.

Section 1. *Be it enacted by the General Assembly of the State of Tennessee*, That it shall be unlawful for any teacher in any of the Universities, Normals and all other public schools of the State which are supported in whole or in part by the public school funds of the State, to teach any theory that denies the story of the Divine Creation of man as taught in the Bible, and to teach instead that man has descended from a lower order of animals.

Section 2. *Be it further enacted*, That any teacher found guilty of the violation of this Act, Shall be guilty of a misdemeanor and upon conviction, shall be fined not less than One Hundred \$ (100.00) Dollars nor more than Five Hundred (\$ 500.00) Dollars for each offense.

Section 3. *Be it further enacted*, That this Act take effect from and after its passage, the public welfare requiring it.

Passed March 13, 1925 (Repealed September 1, 1967)

THE THREE LAWS OF ROBOTICS

The Czech word robota refers to compulsory, or serf, labour. In his 1920 play R. U. R. (Rossum's Universal Robots), the Czech writer and playwright Karel Čapek co-opted it and coined 'robot' to designate a class of automatons designed to work for humans. In what has now become a pretty standard turn of events for stories featuring such creations, they end up rebelling and destroying the human race.

Thirty years after the play was published, the science fiction author Isaac Asimov presented 'the three laws of robotics' in the frontispiece of his novel *I, Robot*. The idea is that if they are followed they should allow the safe coexistence of robots and human beings. They are:

1. A robot may not injure a human being, or, through inaction, allow a human being to come to harm.
2. A robot must obey the orders given it by human beings except where such orders would conflict with the first law.
3. A robot must protect its own existence as long as such protection does not conflict with the first or second Law.

Asimov added a fourth law in later novels, which he named the zeroth law because it was meant to precede those above. Giving it this name also made the laws resonate with those of thermodynamics (see page 81). It stated that:

1. A robot may not harm humanity, or, by inaction, allow humanity to come to harm.

> *An idea isn't responsible for the people who believe in it.*
>
> **Don Marquis, 'The Sun Dial Column',** *New York Sun*, 1918

FIVE, FOUR, THREE, TWO, ONE …
WE HAVE LIFT-OFF

From launch, every space shuttle flight took just nine minutes to attain orbit and a speed of over 28,000km/h (more than 22 times the speed of sound). Helping to achieve this were three very carefully thought out chemical reactions.

The two solid fuel rocket boosters attached to its external fuel tank were chiefly responsible for powering the first two minutes of the shuttle's journey, that's about the first 45km. At the point of lift-off, the total weight of the shuttle was 600 tonnes, so an incredibly large lifting force was needed to start it on its travels. The boosters contained a mixture of the fuel (powdered aluminium), a powerful oxidiser (ammonium perchlorate), a small amount of catalyst (Iron (III) Oxide) to help the reaction along, plus a binder and curing agent. These reacted together to produce steam and nitric oxide that, due to the temperature inside the boosters reaching over 3,000°C from the heat of the reaction, expanded extremely rapidly to create huge propulsive force. Also produced was solid aluminium oxide and aluminium chloride, which could be seen as the dense white clouds expelled from the rockets. In case you are feeling extremely clever, the equation for this reaction is:

$$3Al(s) + 3NH_4ClO_4(s) \rightarrow Al_2O_3(s) + AlCl_3(s) + 6H_2O(g) + 3NO(g)$$

Once the boosters were jettisoned, the main engines on the shuttle itself became responsible for getting it into orbit. For the next six minutes, these used liquid hydrogen and liquid oxygen stored in the external fuel tank. The resulting water vapour was expelled from the engines at almost 10,000km/h. Here's the equation for that reaction:

$$2H_2(l) + O_2(l) \rightarrow 2H_2O(g)$$

The final propulsion system was used to manoeuvre the shuttle once it was in orbit. It was vital it involved a reaction that could be started and stopped exactly as required. This time the fuel was monomethylhydrazine and the oxidiser dinitrogen tetraoxide. This again produced water vapour, but also carbon dioxide. The reaction goes like this:

$$4CH_3HNNH_2(l) + 5N_2O_4(l) \rightarrow 12H_2O(g) + 4CO_2(g) + 9N_2(g)$$

> *The scientist must set in order. Science is built up with facts, as a house is with stones. But a collection of facts is no more a science than a heap of stones is a house.*
> Henri Poincaré, *Science and Hypothesis*

BLOWING HOT AND COLD

The following is a temperature scale in degrees Kelvin, from absolute zero at the bottom right up to the searing heat of the surface of the Sun.

Temperature in Kelvin

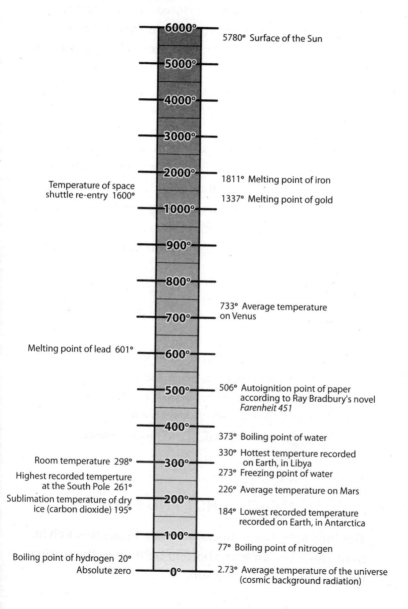

6000° — 5780° Surface of the Sun

5000°

4000°

3000°

2000° — 1811° Melting point of iron

Temperature of space shuttle re-entry 1600° — 1337° Melting point of gold

1000°

900°

800°

700° — 733° Average temperature on Venus

Melting point of lead 601° — 600°

500° — 506° Autoignition point of paper according to Ray Bradbury's novel *Farenheit 451*

400° — 373° Boiling point of water

330° Hottest temperture recorded on Earth, in Libya

Room temperature 298° — 300°

273° Freezing point of water

Highest recorded temperture at the South Pole 261° — 226° Average temperature on Mars

Sublimation temperature of dry ice (carbon dioxide) 195° — 200° — 184° Lowest recorded temperature recorded on Earth, in Antarctica

100°

77° Boiling point of nitrogen

Boiling point of hydrogen 20°

Absolute zero — 0° — 2.73° Average temperature of the universe (cosmic background radiation)

> *A first-rate theory predicts; a second-rate theory forbids and a third-rate theory explains after the event.*
>
> Aleksander Isaakovich Kitaigorodskii,
> crystallographer, at a lecture IUC
> Amsterdam, August 1975

'THERE IS MORE TO SEEING THAN MEETS THE EYEBALL'

Now, what I want is, Facts. Teach these boys and girls nothing but Facts. Facts alone are wanted in life. Plant nothing else, and root out everything else. You can only form the minds of reasoning animals upon Facts: nothing else will ever be of any service to them. This is the principle on which I bring up my own children, and this is the principle on which I bring up these children. Stick to Facts, sir!

Thomas Gradgrind in Charles Dickens' *Hard Times*

Although extreme in his nature, Thomas Gradgrind symbolises perfectly a particular view of science as something cold and clinical involving absolute objectivity. For many years, scientific theories were considered to be something derived purely from facts – crudely put, it was believed that the more facts you had, or examined, the better your theory was likely to be. However, as someone aptly put it, 'the trouble with facts is that there are so many of them' – how are you to decide which of them are relevant? Furthermore, there's an even bigger problem with a fact-leading-to-theory view of science – no matter how hard you try, it's pretty much impossible for observation, typically the source of a scientist's facts, not to be in some way theory-laden. As the philosopher Paul Feyerabend described it 'we interpret our observation statements with the help of the theories we possess [...] and not the other way round'.

This is a subject Norwood Russell Hanson explored in his book *Patterns of Discovery* (1958). Early on he gives the hypothetical example of Johannes Kepler and Tycho Brahe watching the same dawn. Kepler believed the Earth revolved around the Sun while Brahe thought the opposite, that the Earth did not move. So, 'do Kepler and Tycho see the same thing in the east at dawn?' Hanson asked. In terms of sense data, the answer must be an unequivocal yes. But, he argued, 'there is more to seeing than meets the eyeball'. Any observation of something is 'shaped by prior knowledge'. Kepler saw the dawn as the Earth spinning him back to face the Sun while Brahe saw it as the beginning of the Sun's diurnal circuit around the Earth. As Sherlock Holmes critically put it 'one begins to twist facts to suit theories, instead of theories to suit facts'. So, perhaps surprisingly, observed evidence isn't necessarily sufficient for supporting the truth of a theory.

If there are difficulties proving the truth of a theory then what other options does science have? Enter Sir Karl Popper. In his book *The Logic of Scientific Discovery* (German language 1934, English language 1959), he introduced the concept of falsificationism, which states that an attempt should always be made to prove a theory *wrong*. Now the first criterion for a scientific theory or hypothesis to be considered a good one is that it must be falsifiable. This, Popper insisted, was what distinguished science from what he considered to be pseudo-sciences such as Freudian and Marxist theories, which are virtually impossible to falsify (give it a try). The more falsifiable the theory the better because, Popper felt, more was likely to be gleaned from any mistakes. There was little merit in playing it safe.

The idea of falsificationism is undoubtedly a useful one. But it has its limitations. Ironically, one of Popper's favourite scientific examples actually illustrates one of these. In 1919, Sir Arthur Eddington famously set out to an island off West Africa. The purpose of his journey was to test a prediction of Einstein's general theory of relativity regarding the bending of light by gravity through the observation of a solar eclipse. Eddington returned to declare the results as predicted by the theory. So far, so good – here seems the perfect demonstration of how falsificationism should work. We have a theory with big claims and had the results of the expedition been negative then grave doubts would have appeared regarding it.

However, recent historians of science have looked carefully at the data Eddington collected and have come to the conclusion that he was in fact selective with his results. They certainly didn't vindicate Einstein's theory and could even have been presented as falsifying it. And yet the theory is still going strong today precisely because of its excellent predictive power. The problem was that, at heart, the data Eddington recorded was poor – not good enough to prove or disprove the theory. What this shows is that a theory could be falsified through no fault of its own but because of a mistake elsewhere.

The announcement in 2011 of experimental results showing neutrinos travelling faster than light illustrates this further. At the time, reports focused on the challenge this presented to Einstein's special theory of relativity – the results, in effect, falsified it – and much was made of this. The hoo-ha, however, turned out to be due to a hardware error.

CONCEDING TO THE LAWS OF METRE

In his poem 'Vision of Sin' (1842), Alfred Lord Tennyson includes the lines:

> *Every moment dies a man*
> *Every moment one is born.*

These lines prompted some wry editorial comments from Charles Babbage, mathematician and designer of the analytical engine, arguably the first programmable computer. Babbage wrote to the great man commenting, 'I need hardly point out to you that this calculation would tend to keep the sum total of the world's population in a state of perpetual equipoise whereas it is a well-known fact that the said sum total is constantly on the increase. I would therefore take the liberty of suggesting that in the next edition of your excellent poem the erroneous calculation to which I refer should be corrected as follows:

> *Every moment dies a man*
> *And one and a sixteenth is born.*

I may add that the exact figures are 1.167, but something must, of course, be conceded to the laws of metre.'

Needless to say, in the second edition of the poem, published nine years later, these lines stayed exactly as they were.

FIRST, DO NO HARM

The *Hippocratic Corpus* is a collection of about 60 ancient Greek medical tracts written by a number of different authors over a 50 year period during the 4th and 5th centuries B.C. Included within it is the 'Hippocratic Oath', which lays down a physician's obligations. It's a document that has been rewritten throughout history to reflect the outlook of that particular age and culture, which makes reading this translation of it in its original form by the National Library of Medicine's Michael North all the more interesting.

> *I swear by Apollo the physician, and Asclepius, and Hygieia and Panacea and all the gods and goddesses as my witnesses, that, according to my ability and judgement, I will keep this Oath and this contract:*

To hold him who taught me this art equally dear to me as my parents, to be a partner in life with him, and to fulfill his needs when required; to look upon his offspring as equals to my own siblings, and to teach them this art, if they shall wish to learn it, without fee or contract; and that by the set rules, lectures, and every other mode of instruction, I will impart a knowledge of the art to my own sons, and those of my teachers, and to students bound by this contract and having sworn this Oath to the law of medicine, but to no others.

I will use those dietary regimens which will benefit my patients according to my greatest ability and judgement, and I will do no harm or injustice to them.

I will not give a lethal drug to anyone if I am asked, nor will I advise such a plan; and similarly I will not give a woman a pessary to cause an abortion.

In purity and according to divine law will I carry out my life and my art.

I will not use the knife, even upon those suffering from stones, but I will leave this to those who are trained in this craft.

Into whatever homes I go, I will enter them for the benefit of the sick, avoiding any voluntary act of impropriety or corruption, including the seduction of women or men, whether they are free men or slaves.

Whatever I see or hear in the lives of my patients, whether in connection with my professional practice or *not, which ought not to be spoken of outside, I will keep secret, as considering all such things to be private.*

So long as I maintain this Oath faithfully and without corruption, may it be granted to me to partake of life fully and the practice of my art, gaining the respect of all men for all time. However, should I transgress this Oath and violate it, may the opposite be my fate.

THE SIZE AND GRAVITY OF THE SITUATION

Eratosthenes of Cyrene (c. 276–195 B.C.) was aware that at noon on the day of the summer solstice a rod planted in the ground at Syene (modern day Aswan in Egypt) cast no shadow. This meant that the Sun was directly overhead. At exactly the same time, a rod at Alexandria cast a shadow showing the angle between the rod and the direction of the Sun's rays to be 1/50th of a circle. Using some knowledge of geometry, it's easy to see that this angle is equal to the angle between the two rods from the Earth's centre, or 7.2° (1/50th of 360°).

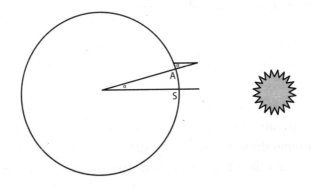

This means that the distance from Syene to Alexandria, which Eratosthenes was able to estimate at 5,000 stades, was 1/50th of the Earth's circumference. A stade – which is one lap of a stadium – is thought to be about 157.5 metres. If this was the length Eratosthenes had in mind, then that makes his calculation of the Earth's circumference over 99% accurate, given that the resultant figure is 39,690km, a little under 400km out from the true figure, which is 40,075km. Even allowing for a more likely stade length of 185 metres, which would give a circumference of 46,250km and an error of just over 16%, Eratosthenes's achievement is still remarkable when we realise he lived at a time when it wasn't even clear that the Earth was spherical.

Of course, his calculation assumes that the Earth is a perfect sphere, which it isn't. Like many of us, our planet exhibits an equatorial bulge and has been described as 'slightly potato shaped'. It's not of uniform density either. Gravitational pull at different locations on a non-uniform mass will vary through being closer or further from the centre of gravity of that mass. This is one of the major reasons why gravity varies by as much as ±0.5% around the world – if you ever want to lose some weight then one piece of advice would be to move to the equator. Sadly while your weight would change your mass would, of course, stay the same, so there'd be little health benefit to this.

In an experiment with echoes of the travelling gnome storyline in the French film *Amelie*, a garden gnome called Kern, who is in the enviable position of having a fixed mass, began blogging in late 2011. He travels round the world and gets weighed at various locations using the same model of electronic scale each time. A true gnome of science, the idea is to investigate precisely how Kern's weight depends on where in the world he is. Here are some of his findings so far:

Location	Weight
The South Pole	309.82g
Under Newton's tree, Woolthorpe, UK	308.59g
Balingen, Germany (Kern's home town)	308.26g
Tokyo, Japan	307.9g
Sydney, Australia	307.8g
CERN, Geneva	307.65g
Mexico City, Mexico	307.62g
Mumbai, India	307.56g

TOP OF THE POP-SCI CHART

(source: Nielsen BookScan)

It's true that the general public's appetite for many subjects changes over time and new trends can appear but popular science has always had a good level of demand and never more so than between 2000 and 2009. Here are the twenty bestselling popular science books of that period, though in the case of series, only the most successful title has been included. Get reading!

1. *A Short History of Nearly Everything*, Bill Bryson – 1,735,617 copies

2. *Why Don't Penguins' Feet Freeze?*, New Scientist – 744,144 copies
3. *Bad Science*, Ben Goldacre – 204,705 copies
4. *A Brief History of Time*, Stephen Hawking – 202,900 copies
5. *The Universe in a Nutshell*, Stephen Hawking – 161,490 copies
6. *The Selfish Gene*, Richard Dawkins – 137,367 copies
7. *Galileo's Daughter*, Dava Sobel – 100,845 copies
8. *The Code Book*, Simon Singh – 98,817 copies
9. *The Elegant Universe*, Brian Greene – 96,619 copies
10. *Longitude*, Dava Sobel – 83,205 copies
11. *The Science of Discworld*, Terry Pratchett, Ian Stewart and Jack Cohen – 74,073 copies
12. *Genome*, Matt Ridley – 72,914 copies
13. *Fermat's Last Theorem*, Simon Singh – 71,010 copies
14. *The Greatest Show on Earth*, Richard Dawkins – 70,736 copies
15. *The Ancestor's Tale*, Richard Dawkins – 63,282 copies
16. *Collapse*, Jared Diamond – 62,144 copies
17. *In Search of Schrodinger's Cat*, John Gribbin – 57,574 copies
18. *Chaos*, James Gleick – 54,939 copies
19. *On the Origin of Species*, Charles Darwin – 44,006 copies
20. *Professor Stewart's Cabinet of Mathematical Curiosities*, Ian Stewart – 41,795 copies

It is by logic that we prove, but by intuition that we discover. To know how to criticise is good, to know how to create is better.

Henri Poincaré, *Science and Method*

A PARTY OF FAMOUS PHYSICISTS

Not dissimilar to the earlier 'How do they do IT' entry (page 7), the following is another popular scientific joke.

> One day, all of the world's famous physicists decided to get together for a tea luncheon. Fortunately, the doorman was a graduate student, and able to observe some of the guests ...

> Everyone gravitated toward Newton, but he just kept moving around at a constant velocity and showed no reaction.
> Einstein thought it was a relatively good time.
> Coulomb got a real charge out of the whole thing.
> Cavendish wasn't invited, but he had the balls to show up anyway.
> Cauchy, being the only mathematician there, still managed to integrate well with everyone.
> Thomson enjoyed the plum pudding.
> Pauli came late, but was mostly excluded from things, so he split.
> Pascal was under too much pressure to enjoy himself.
> Ohm spent most of the time resisting Ampere's opinions on current events.
> Hamilton went to the buffet tables exactly once.
> Volt thought the social had a lot of potential.
> Hilbert was pretty spaced out for most of it.
> Heisenberg may or may not have been there.
> The Curies were there and just glowed the whole time.
> van der Waals forced himself to mingle.

Wien radiated a colourful personality.

Millikan dropped his Italian oil dressing.

de Broglie mostly just stood in the corner and waved.

Hollerith liked the hole idea.

Stefan and Boltzman got into some hot debates.

Everyone was attracted to Tesla's magnetic personality.

Compton was a little scatter-brained at times.

Bohr ate too much and got atomic ache.

Watt turned out to be a powerful speaker.

Hertz went back to the buffet table several times a minute.

Faraday had quite a capacity for food.

Oppenheimer got bombed.

AN UNACKNOWLEDGED DEBT

Western science owes a considerable debt to Islamic scholars, particularly between the 8th and 15th centuries, a period commonly, if misleadingly, referred to as the Dark Ages. It's a time traditionally seen as a low point in the advancement of knowledge, after all the terribly clever things the Greeks and Romans had been up to before that, but actually nothing could be further from the truth. For example, during this period Musa al-Khwarizmi, born around 786 A.D., developed algebra. It was also during this time that the great translation movement took place, a huge intellectual project that saw a mass of texts in ancient Greek and other languages translated into Arabic – this would later enable these ideas to flow back into Europe. And the physician ibn-Sina (his latinised name is Avicenna) wrote his multi-volume *al-Qann fi al-Tibb* (*The Canon of Medicine*) around the turn of the millennium, which became one of the standard medical textbooks in Europe until late in the 17th century.

Scholars today are constantly discovering new connections between the Islamic world of that period and some of the 'breakthroughs' in science that took place elsewhere just afterwards. One of the most interesting of these has been the gradual understanding that Copernicus probably used mathematical models first developed by Islamic astronomers in his proposal that the Earth revolved around the Sun.

Copernicus did acknowledge a debt to Islamic thinking in his ground-breaking book *De revolutionibus*, although there was nothing specifically mentioned that dates to later than around 1200. However, we now know that within Islamic astronomy, there was already a wealth of mathematically based criticism of the Ptolemaic system which Copernicus' system was effectively overthrowing. As there was no tradition in Europe of questioning Ptolemy, there is a growing conviction that some of this work directly influenced Copernicus. Some tangible examples of possible influence given by Professor F. Jamil Regep of McGill University include:

- Probable use of the Tusi couple, a method developed by Nasir al-Din al-Tusi during the 13th century to show how the appearance of movement in a straight line can be produced by a circle rolling inside another circle twice its radius (it's worth watching one of the animations of this on in the internet to see how exactly this works). This enabled Copernicus to deftly sidestep a feature of Ptolemy's system. The lettering of a relevant diagram in *De revolutionibus* follows the customary Arabic lettering as opposed to the Latin, despite the book being written in the latter.

- There's a striking similarity between a section of *De revolutionibus* and one in al-Tusi's *Tadhkira*, which appeared in many subsequent Islamic works.

- There are astronomical models featured in the work of ibn al-Shatir, who worked in Damascus in the 14th century, which, until their discovery, had been previously thought to have originated with Copernicus. The implication is that Copernicus had seen them and used them in his own work.

ELEMENTS OF COLOUR

The characteristic colours of fireworks are due to metals, and their salts, being added to the gunpowder. Particular hues are due to particular metals and their salts – exactly the same ones are seen in laboratory flame tests. This is because when a metal salt is heated, such as in a flame or the ignited gunpowder, the electrons in the metal atoms get promoted to higher energy levels. When the electrons return to their normal, or ground, state, energy is emitted as light at wavelengths specific to that metal. In the case of our fireworks these frequencies lie in the visible region, and it is the frequency that determines the colour we then see.

Metal, or salt of	Colour
Magnesium/Aluminium/Titanium	Intense white
Sodium	Yellow
Strontium/Lithium	Red
Barium	Green
A mixture of Strontium and Copper	Purple
Copper	Blue
Calcium	Orange

The same principle lies behind the magnificent colours of the aurora borealis and aurora australis (northern and southern lights). Electrons in atoms and molecules high in the Earth's atmosphere, such as nitrogen and oxygen, are promoted to higher energy levels by charged particles from the Sun. As they fall back to their ground state, the swirling, heavenly lights are produced.

Police arrested two kids yesterday – one was drinking battery acid, the other was eating fireworks. They charged one and let the other one off.

Tommy Cooper, comedian

CONGRATULATIONS, YOU'VE WON A ...

Originally proposed by Steve Selvin in a letter to *American Statistician*, and inspired by the US TV gameshow *Let's Make a Deal* hosted by Monty Hall, this probability puzzle is appropriately named the Monty Hall problem.

You are on a game show and have got through the last round. The grand prize is almost within your grasp. The host presents you with three doors numbered 1–3. Behind one of these is a luxury car; behind the other two are goats. You choose a door. Before it is opened, the host, who knows what's behind each door, opens one of the others, revealing one of the goats. He then asks you if you want to change your mind from your original choice to the other unopened door. The question is whether this would be to your advantage.

Instinctively we all feel that this would make absolutely no difference at all, but in fact the answer is yes, absolutely you should switch. If you change your choice your odds of winning the car will become 2/3 whereas if you stick with your original choice, it'll be 1/3. Don't believe me? Here's how it works:

Let's assume you picked door 1. We can now show the possible arrangements of the prizes (if a goat is a prize).

Possible prize arrangements	Door 1	Door 2	Door 3
A	Car	Goat	Goat
B	Goat	Car	Goat
C	Goat	Goat	Car

So, if your initial choice of door has a goat behind it (if, say, you picked door 1 and there was a goat, meaning the set up was either arrangement B or C) then switching your choice will definitely win you the car because Monty Hall would open the other door with a goat behind it. In two out of the three possible arrangements, switching choice would win you the car. Sticking with your choice would win you the car in only one of the three arrangements. If we take arrangement A as an example, if you picked door 1 then you'd win by sticking. But if you picked either door 2 or door 3, then you'd win by switching. You see?

A version of the problem was printed in *Parade* magazine in 1990 and prompted nearly 10,000 complaints that the magazine had got it wrong and that there was no benefit to the switch.

> *He uses statistics as a drunken man uses lampposts – for support rather than for illumination.*
>
> **Andrew Lang, Victorian intellectual and writer**

THE TEN GREATEST EVER EQUATIONS ... ACCORDING TO NICARAGUA

In 1971, the Nicaraguan postal service issued a series of postage stamps depicting the ten mathematical equations that, in their opinion, had most changed the world. Each stamp included on its reverse a justification of its selection. Here they are, translated:

1 + 1 = 2
Primitive Man

Simple as it is, this equation had enormous consequences for humanity because it formed the basis of counting. Without an understanding of numbers people could only trade in a rudimentary way; they didn't have an exact tally of the number of sheep or cows or how many men made up the tribe. The discovery led directly to the rapid development of trade and later to the important science of measurement.

$A^2 + B^2 = C^2$
Pythagoras 570–497/6 BC

The most frequently used theorem in geometry is undoubtedly that of Pythagoras which refers to the lengths of the three sides, a, b and c of a right angle triangle. It provided for the first time a means of calculating lengths by indirect means, thereby allowing man to do surveying and produce maps. The ancient Greeks used it to measure such things as the distances of ships at sea and the heights of buildings. Today scientists and mathematicians use it constantly.

$F_1 x_1 = F_2 x_2$
Archimedes 281–212 BC

Archimedes said, 'Give me a place to stand and I will move the world'. The simple equation of the lever is the basis of all engineering, whether it be merely with a crowbar, or the most advanced gearing or crane. It is essential for the design of machinery and all structures from bridges to buildings. Every nut and bolt employs the principle. Our car brakes, door handles, scales, and most tools are varieties of the lever.

$$e^{\ln N} = N$$

John Napier 1550–1617

With the invention of logarithms, Napier gave the world a powerful shorthand for arithmetic. It enabled men to do multiplication or division by simply adding or subtracting the logarithms of numbers, which allowed them to carry these operations out faster, as well as more complicated ones containing many numbers. The impact of logarithms in fields such as astronomy and navigation was huge and comparable to the modern era's computer revolution.

$$f = Gm_1m_2/r^2$$

Sir Isaac Newton 1642–1727

Before Newton's time, people had little idea of the force that kept the planets in their orbits around the Sun, or the Moon around the Earth – or even what stopped us from flying off the surface of the Earth into space. Newton showed that the force of gravity attracts all bodies to one another. The equation shows, however, that although the force depends on the masses of the bodies, we do not notice it between everyday objects because it is comparatively very weak.

$$\nabla^2 E = \frac{ku}{c^2} \frac{\delta^2 E}{\delta t^2}$$

James Clerk Maxwell 1831–1879

A century ago this Scottish physicist discovered four famous equations summarising man's knowledge of electricity and magnetism. From these he then obtained this one equation, along with another predicting the possibility of radio waves. We owe Maxwell all our TV and radio broadcasting, our long distance communication and radar on land, at sea and in space. Light, X-rays and other electromagnetic radiation are also governed by this fundamental equation.

$$S = k \log W$$

Ludwig Boltzmann 1844–1906

Boltzmann's equation revealed how the behaviour of gases depended on the constant motion of atoms and molecules. Its great importance lies in its application in areas in which gases play an important role: in all machines driven by steam or internal combustion, in gas reactions of chemicals used to make modern drugs, plastics or other substances; in understanding weather; and even to explain the violent processes of the Sun, stars and distant galaxies.

$V = V_e \ln(m_0/m_1)$

Konstantin Tsiolkovsky 1857–1935

A basic part of space technology, this equation gives the changing velocity of a ship as it burns the fuels it's carrying. The equation is derived directly from one of the three laws of motion of Sir Isaac Newton. Without it, launching spacecraft to the Moon and planets or orbiting the Earth would be impossible, and it has also made practicable the use of guided rockets in warfare.

$E = mc^2$

Albert Einstein 1875–1955

This equation lies at the root of our nuclear age. It simply says that a small quantity of matter can be converted into a large amount of energy. We see this nuclear energy released in a spectacular and violent way in atomic and hydrogen bombs. But man has also managed to tame 'nuclear fission' in reactors that supply heat and generate electricity for our homes and factories.

$\lambda = h/mv$

Louis de Broglie 1892–1977

Light, a form of energy, can behave both as bullet-like particles and as a continuous wave. De Broglie discovered the converse, that the elementary particles of which matter is composed also have properties that resemble waves. His equation has had a large effect on physics, leading to modern optics, electronic components – for example transistors – with many applications in radio, TV, computers, spacecraft, and military weapons. It also provided scientists with the powerful electron microscope.

> *It is of the highest importance in the art of detection to be able to recognise out of a number of facts which are incidental and which vital.*
>
> Sherlock Holmes in 'The Reigate Puzzle'

IS HELL EXOTHERMIC OR ENDOTHERMIC?

Perhaps inspired by the earlier entry on Hell (see page 89), the following seems to have begun its life on the internet sometime in 1997 and multiple examples now exist. Its opening often varies – who the setter supposedly was and at what institution – but the core of the piece is always the same. Enjoy!

*

The following is an actual question given on a University of Washington engineering mid-term paper. The answer was so profound that the Professor shared it with colleagues, which is why we now have the pleasure of enjoying it as well.

Bonus question: Is Hell exothermic (gives off heat) or endothermic (absorbs heat)?

Most of the students wrote proofs of their beliefs using Boyle's Law (gas cools off when it expands and heats up when it is compressed) or some variant. One student, however, wrote the following:

'First, we need to know how the mass of Hell is changing over time. So we need to know the rate that souls are moving into Hell and the rate they are leaving. I think that we can safely assume that once a soul gets to Hell, it will not leave. Therefore, no souls are leaving. As for how many souls are entering Hell, let's look at the different religions that exist in the world today. Some of these religions state that if you are not a member of their religion, you will go to Hell. Since there are more than one of these religions and since people do not belong to more than one religion, we can project that all souls go to Hell. With birth and death rates as they are, we can expect the number of souls in Hell to increase exponentially.

Now, we look at the rate of change of the volume in Hell because Boyle's Law states that in order for the temperature and pressure in Hell to stay the same, the volume of Hell has to expand as souls are added. This gives two possibilities:

1. If Hell is expanding at a slower rate than the rate at which souls enter Hell, then the temperature and pressure in Hell will increase until all Hell breaks loose.

2. Of course, if Hell is expanding at a rate faster than the increase of souls in Hell, then the temperature and pressure will drop until Hell freezes over.

So which is it? If we accept the postulate given to me by Teresa Banyan during my Freshman year, that "… it will be a cold day in Hell before I sleep with you" and take into account the fact that I still have not succeeded in having sexual relations with her, then, #2 cannot be true, and thus I am sure that Hell is exothermic and will not freeze.'

Lise Meitner, 1878–1968
A physicist who never lost her humanity

Gravestone, St. James's Church,
Bramley, Hampshire, UK

GREAT SCIENTISTS BY NATION

In 2002 the BBC ran a poll to discover the 100 Greatest Britons, following it up with a series of programmes on the subject. The format was subsequently copied by a number of other countries who in turn set out to discover their 100 Greatest Americans, 100 Greatest Belgians etc. Some curious quirks cropped up: the Canadian chart featured a number of people who weren't technically Canadian, such as Alexander Graham Bell (born in Scotland but later took up US citizenship); Haroun Tazieff did study in Belgium, and lived there for a period of time, but is considered a Polish-born Frenchman; similarly Marie Curie was Polish and named the element Polonium after her home country but featured in France's list and not Poland's; and Albert Einstein appeared in both the German and American lists. The following gives the top 5 scientists from a number of these polls plus their overall positions in brackets.

100 Greatest Britons (as conducted by the BBC)

1. Charles Darwin (1809–1882) (4), see page 56.
2. Sir Isaac Newton (1642–1727) (6), see page 207.
3. Sir Alexander Fleming (1881–1955) (20), winner of the Nobel Prize in Physiology or Medicine, 1945, with Ernst B. Chain and Sir Howard Florey 'for the discovery of penicillin and its curative effect in various infectious diseases'.
4. Alan Turing (1912–1954) (21), mathematician and founding father of computer science.
5. Michael Faraday (1791–1867) (22), see page 92.

Le Plus Grand Belge, Belgium (as conducted on French-speaking public channel RTBF)

1. Ernest Solvay (1838–1922) (15), industrial chemist, benefactor and promoter of scientific research and popularisation.
2. Andreas Vesalius (1514–1564) (19), anatomist and author of *De humani corporis fabrica* (1543), which created a revolution in the understanding of human anatomy.
3. Jules Bordet (1870–1961) (37), winner of the Nobel

Prize in Physiology or Medicine, 1919, 'for his discoveries relating to immunity'.

4. Haroun Tazieff (1914–1998) (46), volcanologist who communicated his subject through many books and dramatic films.

5. Ilya Prigogine (1917–2003) (72), winner of the Nobel Prize in Chemistry, 1977, 'for his contributions to non-equilibrium thermodynamics'.

The Greatest Canadian (conducted by CBC)

1. Frederick Banting (1891–1941) (4), joint winner with J. J. R. Macleod of the Nobel Prize in Physiology or Medicine, 1919, 'for the discovery of insulin'.

2. Alexander Graham Bell (1844–1922) (9), inventor of the telephone.

3. Sir Sandford Fleming (1827–1915) (42), engineer and inventor of Universal Standard Time, which divided the globe into 24 time zones.

4. William Edmond Logan (1798–1875) (61), geologist who worked out how coal is formed and founder of the Geological Survey of Canada.

5. Charles Best (1899–1978) (77), co-discoverer of insulin who was overlooked for the Nobel Prize mentioned above.

Největší Čech, The Czech Republic (conducted by Česká Televise)

1. Jaroslav Heyrovský (1890–1967) (19), winner of the Nobel Prize in Chemistry, 1959, 'for his discovery and development of the polarographic methods of analysis'.

2. Otto Wichterle (1913–1998) (23), chemist and inventor of soft contact lenses.
3. Jan Evangelista Purkyně (1787–1869) (40), physiologist who made countless discoveries including that of the uniqueness of fingerprints.
4. Jan Janský (1873–1921) (42), physician who identified the four blood groups A, B, AB and O.
5. Gregor Mendel (1822–1884) (54), father of genetics.

Le Plus Grand Français de Tous les Temps, France (conducted by France 2)

1. Louis Pasteur (1822–1895) (2), see page 187.
2. Marie Curie (1867–1934) (4), the first person to win two Nobel Prizes (Physics and Chemistry).
3. Jacques Cousteau (1910–1997) (9), underwater explorer, conservationist and film-maker.
4. Christian Cabrol (b. 1925) (37), surgeon and pioneer of heart and lung transplantation.
5. Haroun Tazieff* (47).

* see also the Greatest Belgian list

Unsere Besten, Germany (conducted by ZDF)

1. Albert Einstein (1879–1955) (10), see page 163.
2. Robert Bosch (1861–1942) (14), engineer, inventor and founder of the *Bosch* company.
3. Konrad Zuse (1910–1995) (16), inventor of the first programmable computer.
4. Wilhelm Conrad Röntgen (1845–1923) (28), see page 158.
5. Robert Koch (1843–1910) (51), see page 187.

De Grooste Nederlander, the Netherlands (conducted by KRO)

1. Antonie van Leeuwenhoek (1632–1723) (4), the first person to observe bacteria.
2. Christiaan Huygens (1629–1695) (12), physicist and astronomer who asserted light was a wave and discovered Saturn's rings.
3. Hendrik Lorentz (1853–1928) (49), winner of the Nobel Prize in Physics, 1902, with Pieter Zeeman 'in recognition of the extraordinary service they rendered by their researches into the influence of magnetism upon radiation phenomena'.
4. Johannes Diderik van der Waals (1837–1923) (74), winner of the Nobel Prize in Physics 1910 'for his work on the equation of state for gases and liquids'.
5. Herman Boerhaave (1668–1738) (77), botanist and physician who wrote one of the first textbooks of physiology.

New Zealand's Top 100 History Makers (conducted by Prime)

1. Ernest Rutherford (1871–1937) (1), winner of the Nobel Prize in Chemistry, 1908, 'for his investigations into the disintegration of the elements, and the chemistry of radioactive substances'.
2. Sir Brian Barratt-Boyes (1924–2006) (11), second person to perform a heart-valve replacement.
3. William Pickering (1910–2004) (13), pioneering space scientist and former director of the Jet Propulsion Laboratory in the US.

4. Maurice Wilkins (1916–2004) (26), winner of the Nobel Prize in Physiology or Medicine, 1962, with Francis Crick and James Watson 'for their discoveries concerning the molecular structure of [DNA] and its significance for information transfer in living material'.
5. Sir Harold Gillies (1882–1960) (30), pioneer of plastic surgery.

Greatest American (conducted by AOL and the Discovery Channel)
1. Benjamin Franklin (1706–1790) (5), statesman and polymath who made significant contributions in the science of electricity.
2. Albert Einstein (1879–1955) (14), see page 163.
3. Thomas Edison (1847–1931) (15), responsible for many inventions including the incandescent light bulb.

The next final 75 names weren't ranked but included Alexander Graham Bell, George Washington Carver, Carl Sagan, and Nikola Tesla.

The poll for the Top 100 Historical Persons in Japan wasn't actually restricted to the Japanese and included many people from other countries:

1. Thomas Edison (1847–1931) (3).
2. Hideyo Noguchi (1876–1928) (7), bacteriologist who discovered the cause of syphilis and died of yellow fever while researching the disease.
3. Albert Einstein (1879–1955) (13), see page 163.

4. Florence Nightingale (17), reformed nursing and was the first woman to be elected a Fellow of the Royal Statistical Society.

5. Leonardo da Vinci (1452–1519) (23), polymath who embodies the idea of the Renaissance Man.

> *Intended for Sir Isaac Newton*
>
> *Nature and Nature's laws lay hid in night.*
> *God said, Let Newton be! And all was light.*
>
> Alexander Pope

'SUCCESSIVE GENERATIONS BLOOM'

Nearly 60 years before his grandson Charles's *On the Origin of Species*, Erasmus Darwin (1731–1802) wrote *The Temple of Nature, or, The Origin of Society*, a poetic hymn to evolution, which in successive cantos discussed the 'production of life', the 'reproduction of life', 'progress of the mind' and 'good and evil'. A member of the Lunar Society in Birmingham, which also included the steam engine partners, James Watt and Matthew Boulton, Joseph Priestley, most famous for his discovery of oxygen and the innovative potter Josiah Wedgwood, Darwin's final work sought 'simply to amuse by bringing distinctly to the imagination the beautiful and sublime images of the operations of Nature in the order [...] in which the progressive course of time presented them'.

From Canto I of The Temple of Nature

Organic Life beneath the shoreless waves
Was born and nurs'd in Ocean's pearly caves;
First forms minute, unseen by spheric glass,
Move on the mud, or pierce the watery mass;
These, as successive generations bloom,
New powers acquire, and larger limbs assume;
Whence countless groups of vegetation spring,
And breathing realms of fin, and feet, and wing.

Thus the tall Oak, the giant of the wood,
Which bears Britannia's thunders on the flood;
The Whale, unmeasured monster of the main,
The lordly Lion, monarch of the plain,
The Eagle soaring in the realms of air,
Whose eye undazzled drinks the solar glare,
Imperious man, who rules the bestial crowd,
Of language, reason, and reflection proud,
With brow erect who scorns this earthy sod,
And styles himself the image of his God;
Arose from rudiments of form and sense,
An embryon point, or microscopic ens!

Success in research needs four Gs:
Glück, Geduld, Geschick und Geld.
[Luck, patience, skill and money]

Paul Ehrlich as quoted in M. Perutz, 'Rita and the Four Gs', *Nature*, 338, 791 (1988)

IT HAS BEEN LONG KNOWN THAT ...

Have you ever wondered what certain phrases in scientific papers really mean? Well, in 1957 C. D. Graham, who worked in a General Electric Company research laboratory, lifted the lid when he wrote the following 'glossary for research reports' in the journal *Metal Progress*.

*

INTRODUCTION	
It has been long known that...	I haven't bothered to look up the original reference
... of great theoretical and practical importance	...interesting to me
While it has not been possible to provide definite answers to these questions ...	The experiments didn't work out, but I figured I could at least get a publication out of it

EXPERIMENTAL PROCEDURE	
The W-Pb system was chosen as especially suitable to show the predicted behaviour ...	The fellow in the next lab had some already made up
High purity ... Very high purity ... Extremely high purity ... Super-purity ... Spectrascopically pure...	Composition unknown except for the exaggerated claims of the supplier
A fiducial reference line ...	A scratch
Three of the samples were chosen for detailed study ...	The results on the others didn't make sense and were ignored
... accidentally strained during mounting	... dropped on the floor
... handled with extreme care throughout the experiments	... not dropped on the floor

RESULTS	
Typical results are shown ...	The best results are shown
Although some detail has been lost in reproduction, it is clear from the original micrograph that ...	It is impossible to tell from the micrograph

RESULTS (continued)

Presumably longer times ...	I didn't take time to find out
The agreement with the predicted curve is excellent	Fair
Good	Poor
Satisfactory	Doubtful
Fair	Imaginary
... as good as could be expected	Non-existent
These results will be reported at a later date	I might possibly get around to this sometime
The most reliable values are those of Jones	He was a student of mine

DISCUSSION

It is suggested that ... It is believed that ... It may be that ...	I think
It is generally believed that ...	A couple of other guys think so too
It might be argued that ...	I have such a good answer to this objection that I shall raise it now
It is clear that much additional work will be required before a complete understanding ...	I don't understand it
Unfortunately, a quantitative theory to account for these effects has not been formulated	Neither does anyone else
Correct within an order of magnitude	Wrong
It is hoped that this work will stimulate further work in the field	This paper isn't very good, but neither are any of the others in this miserable subject

ACKNOWLEDGEMENTS

Thanks are due to Joe Glotz for assistance with the experiments and to John Doe for valuable discussions	Glotz did the work and Doe explained what it meant

WHAT IF EARTH CAN CLOTHE AND FEED AMPLEST MILLIONS AT THEIR NEED

The radical poet Percy Bysshe Shelley fought against the system throughout his short life. At Eton, he wouldn't 'fag' – a system where a younger boy effectively serves an older one – and was ostracised by his fellow pupils. At Oxford, he lasted only a term after writing and then refusing to withdraw *The Necessity of Atheism*, a pamphlet he co-wrote with his college friend, Thomas Jefferson Hogg, and which had been sent to every head of college.

A vociferous and varied reader, he was a thinker of extraordinary ideas and it's hard to find a poet more enamoured with science and the promise it held for improving the lot of the general populace. Shelley's notes on Humphry Davy's *Elements of Agricultural Chemistry* alone ran to over twenty pages and he possessed all manner of scientific equipment in his college rooms including an electrical machine and an air pump.

Hogg wrote about their time at Oxford together in *The New Monthly Magazine and Literary Journal* just over twenty years later, a decade after Shelley's premature death. In it is a passage that relates Shelley's belief that science could benefit society, improving the welfare of everyone, particularly the impoverished. Even taking into account the second hand nature of the account, and while still trying to be careful about reading predictions made 200 years ago with today's eyes, Shelley's thoughts are nevertheless remarkable. Allowing for the odd tweak here and there, one could argue that nearly every single one has come true. It's the optimism science can provide, as articulated by Shelley, that makes it as important and as wonderful as it is. It just needs to be handled in the way Shelley imagines.

*

'Is not the time of by far the larger proportion of the human species', he inquired, with his fervid manner and in his piercing tones, 'wholly consumed in severe labour? And is not this devotion of ... the whole of our race (for those who ... are indulged with an exemption from the hard lot are so few ... that they scarcely deserve to be taken into the account) absolutely necessary to procure subsistence; so that men have no leisure for recreation or the high improvement of the mind? Yet this incessant toil is still inadequate to procure an abundant supply of the common necessaries of life: some are doomed actually to want them, and many are compelled to be content with an insufficient provision.

'We know little of the peculiar nature of those substances which are proper for the nourishment of animals; we are ignorant of the qualities that make them fit for this end. Analysis has advanced so rapidly of late that we may confidently anticipate that we shall soon discover wherein their aptitude really consists; having ascertained the cause, we shall next be able to command it, and to produce at our pleasure the desired effects.

'It is easy, even in our present state of ignorance, to reduce our ordinary food to carbon, or to lime; a moderate advancement in chemical science will speedily enable us, we may hope, to create, with equal facility, food from substances that appear at present to be as ill adapted to sustain us.

'What is the cause of the remarkable fertility of some lands, and of the hopeless sterility of others? A spadeful of the most productive soil does not to the eye differ much from the same quantity taken from the most barren. The real difference is probably very slight; by chemical agency the philosopher may work a total change, and may transmute an unfruitful region into a land of exuberant plenty.

'Water, like the atmospheric air, is compounded of certain gases: in the progress of scientific discovery a simple and sure method of manufacturing the useful fluid, in every situation and in any quantity, may be detected; the arid deserts of Africa may then be refreshed by a copious supply, and may be transformed at once into rich meadows, and vast fields of maize and rice.

'The generation of heat is a mystery, but enough of the theory of caloric has already been developed to induce us to acquiesce in the notion that it will hereafter, and perhaps at no very distant period, be possible to produce heat at will, and to warm the most ungenial climates as readily as we now raise the temperature of our apartments to whatever degree we may deem agreeable or salutary. If, however, it be too much to anticipate that we shall ever become sufficiently skilful to command such a prodigious supply of heat, we may expect, without the fear of disappointment, soon to understand its nature and the causes of combustion, so far at least as to provide ourselves cheaply with a fund of heat that will supersede our costly and inconvenient fuel, and will suffice to warm our habitations, for culinary purposes and for the various demands of the mechanical arts.

'We could not determine, without actual experiment, whether an unknown substance were combustible; when we shall have thoroughly investigated the properties of fire, it may be that we shall be qualified to communicate to clay, to stones, and to water itself, a chemical recomposition that will render them as inflammable as wood, coals, and oil, for the difference of structure is minute and invisible, and the power of feeding flame may perhaps be easily added to any substance, or taken away from it. What a comfort would it be to the poor at all times, and especially at this season, if we were capable of solving this problem alone, if we could furnish them with a competent supply of heat!

'These speculations may appear wild, and it may seen improbable that they will ever be realised, to persons who have not extended their views of what is practicable by closely watching science in its course onward; but there are many mysterious powers, many irresistible agents, with the existence and with some of the phenomena of which all are acquainted. What a mighty instrument would electricity be in the hands of him who knew how to wield it, in what manner to direct its omnipotent energies; and we may command an indefinite quantity of the fluid: by means of electrical kites we may draw down the lightning from heaven! What a terrible organ would the supernal shock prove, if we were able to guide it; how many of the secrets of nature would such a stupendous force unlock! The galvanic battery is a new engine; it has been used hitherto to an insignificant extent, yet has it wrought wonders already; what will not an extraordinary combination of troughs, of colossal magnitude, a well-arranged system of hundreds of metallic plates, effect? The balloon has not yet received the perfection of which it is surely capable; the art of navigating the air is in its first and most helpless infancy; the aerial mariner still swims on bladders, and has not mounted even the rude raft: if we weigh this invention, curious as it is, with some of the subjects I have mentioned, it will seem trifling, no doubt a mere toy, a feather, in comparison with the splendid anticipations of the philosophical chemist; yet it ought not altogether to be contemned. It promises prodigious facilities for locomotion, and will enable us to traverse vast tracts with ease and rapidity, and to explore unknown countries without difficulty. Why are we still so ignorant of the interior of Africa? Why do we not despatch intrepid aeronauts to cross it in every direction, and to survey the whole peninsula in a few weeks? The shadow of the first balloon, which a vertical sun would project precisely underneath it, as it glided silently over that hitherto unhappy country, would virtually emancipate every slave, and would annihilate slavery for ever.'

APPENDIX: BACK TO SCHOOL

DOUBLE SCIENCE

School science – did it fill you with dread or delight? For me, that rather depended on the teacher and the area being covered (I'll let you in on a secret; I never got on with electric circuits, which is probably why you won't find anything to do with them in this book).

But school science is the grounding of most, if not all, of our scientific literacy, so it doesn't do any harm to go over some of the things you may have covered there.

We'll start with one of my favourites (don't get cornered by me at a party – seriously), the atom.

CHEMISTRY IS FULL OF BAD JOKES BECAUSE ALL THE BEST ARGON

If you search the depths of your memory, you may recall that the periodic table contains the 118 elements from which *everything* is made and that the smallest unit of any element is the atom. Atoms themselves are made up of three sub-atomic particles: protons, neutrons and electrons (you'll discover in the Standard Model entry, that it's actually a whole lot more complicated than this, but let's not overcomplicate things for now). The nucleus of an atom, found at its centre, is made up of the neutrons and protons clustered together. Only these contribute to the element's atomic mass. Both neutrons and protons have roughly the same mass. The electrons, on the other hand, have a negligible mass in comparison to the nucleus, which they 'orbit'.

The elements in the periodic table are arranged in order of increasing atomic number, which is determined by the number of protons – all the atoms of any given element always have the same number of protons. The atomic number of carbon, for example, is 6, which means there are 6 protons in every atom of carbon.

While every atom of an element has the same number of protons, the same is not always true of neutrons. Isotopes are atoms of the same element that differ in the number of neutrons they contain. For example, there are three naturally occurring isotopes of carbon: carbon-12, carbon-13 and carbon-14 (the number after the dash refers to the atomic mass of the isotope; so, given that carbon always contains 6 protons, carbon-13 must contain 7 neutrons).

As we just saw, electrons orbit the nucleus of an atom, where its protons and neutrons are packed together. Protons have a relative charge of +1 and electrons have a relative charge of –1 (neutrons are uncharged). So, having an equal number of protons and electrons means that the atom is electrically neutral overall. For example, a neutrally-charged atom of carbon will contain 6 protons hence its atomic number is 6 – and 6 electrons.

Electrons orbiting a nucleus do so according to clearly defined regions called electron shells. Each shell, which corresponds to a level of electron energy, can contain only a fixed number of electrons. The first one, which is the lowest-energy shell, can hold a maximum of 2 electrons only. Each subsequent shell of the first twenty elements of the periodic table is capable of containing 8 electrons. As you move along the periodic table in order of increasing atomic number, these shells are filled up with electrons in sequence. So, hydrogen (the first element in the periodic table) has 1 electron in the first shell and helium has 2. Now the first shell is full, the next element, lithium, will have 2 electrons in the first shell and 1 electron in the second shell. Once we get to neon (with the atomic number 10), both the first shell (2 electrons) and second shell (8 electrons) are now full, so the next element, sodium (with atomic number 11), has a single electron in the third shell. The outermost shell of an atom is called the valence shell.

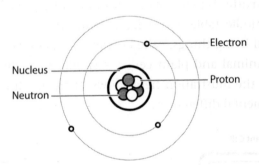

The structure of an atom

If an atom somehow loses or gains an electron, meaning that there are more protons than electrons, or vice versa, it becomes an ion. If an electron is lost, a positively-charged ion (cation) is formed; if an electron is gained, a negatively-charged ion (anion) is formed.

THE SCIENCE MAGPIE

Metals will generally lose electrons to form cations while non-metals will gain electrons to form anions. This can happen when metals and non-metals react with each other, such as when sodium and chlorine combine to form sodium chloride (common table salt). Here, a chlorine atom effectively takes an electron from a sodium atom and an ionic bond is formed between them. One useful way of remembering the difference between cations and anions is the phrase 'cations are pussy-tive'. A chemical ion is denoted by adding the relevant charge after the element's symbol, in superscript, for example: Li^+, Ca^{2+}, F^- and O^{2-}.

CELL-U-LIKE

So, all matter is made of one or more of the elements of the periodic table. On a larger (but still microscopic) scale all living things, called organisms, are made up of cells. Animal and plant cells are similar in many ways but, as the illustration here shows, there are also some fundamental differences.

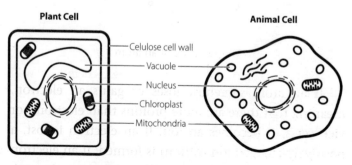

Plant Cell

Animal Cell

Celulose cell wall

Vacuole

Nucleus

Chloroplast

Mitochondria

A plant cell and an animal cell

The features common to both types of cell are:

- The cell membrane is a very thin, porous wall that enables necessary gases and chemical nutrients to pass into and out of the cell.

- The nucleus is effectively the 'control room' of the cell. It also contains the cell's genetic material.

- Both cell types possess a fluid environment outside the nucleus in which everything else in the cell resides; it's called cytoplasm.

- In plant cells there are large, sack-like areas in the cell called vacuoles. They contain high concentrations of nutrients and also assist in the disposal of waste. Not all animal cells possess vacuoles but in those that do they are much smaller in size and more numerous than in plant cells.

- Mitochondria are the 'powerhouses' of both cell types, since they produce the molecule adenosine triphosphate (ATP), the currency of energy in all living organisms.

The differences between plant and animals cells are significant, however:

- Only the plant cell contains a rigid, cellulose wall. This is what gives plants their great strength.

- Only plant cells contain chloroplasts, structures which in turn contain chlorophyll. They give plants their green colour, and are responsible for photosynthesis, the method by which plants turn carbon dioxide and water into sugars such as glucose to provide themselves with energy.

WHO DO YOU THINK YOU ARE?

The concept of the gene, the basic unit of inheritance, has been understood for over a century. But it wasn't until James Watson and Francis Crick, thanks in part to Rosalind Franklin, cracked the structure of Deoxyribonucleic acid (DNA) that the study of genetics really began to take off.

DNA is the molecule in which all our genes are located and it can be found in all our bodies (a gene is a section of DNA – they vary in size). Its now famous structure is a double-helix – a little like a ladder twisted into a spiral.

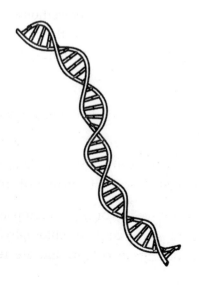

Although the information in the DNA of an organism determines most of what that thing is, it can't predict everything precisely down to the last detail. It's probably better to think of DNA as a recipe or even a script to be interpreted rather than taking the common view of it being a blueprint.

> *Eight hundred life spans can bridge more than 50,000 years. But of these 800 people, 650 spent their lives in caves or worse; only the last 70 had any truly effective means of communicating with one another, only the last 6 ever saw a printed word or had any real means of measuring heat or cold, only the last 4 could measure time with any precision; only the last 2 used an electric motor; and the vast majority of the items that make up our material world were developed within the lifespan of the 800th person.*
>
> Assessing Technology Transfer
> (NASA Report SP-5067), 1966, pp 9–10

In both plants and animals, DNA is packaged up (along with other materials) in the form of chromosomes, which look like two thread-like structures pinched together at a point along their length. There are 46 chromosomes in 23 pairs in every cell of the human body. One of these pairs will be made of two X chromosomes if you're female and an X and Y chromosome if you're male (only men carry a Y chromosome). The disorder Down's syndrome is due to a person having an extra copy of chromosome 21.

When sperm and egg cells are formed, only one chromosome from each pair ends up being in each sperm or egg cell. So, in every egg cell there will be an X chromosome (either one from the female's XX pair) whereas half of the sperm cells will contain an X chromosome and the other half will contain a Y one (either one from the male's XY pair). When a sperm and egg cell fuse, the number of chromosomes is restored to 46, with half of the genetic material coming from the father and the other half from the mother. Whether the sperm cell involved contains an X or a Y chromosome will ultimately determine the baby's sex.

BY A FORCE OF NATURE

We'll be seeing quite a bit of Sir Isaac Newton (1642–1727) in this book. He is famous for many things but at school you are most likely to have come across his three laws. This is how Newton himself expressed them:

First law
Every body perseveres in its state of rest, or of uniform motion in a right line, unless it is compelled to change that state by forces impressed thereon.
The first part of this statement introduces the concept of inertia. The second part goes on to tell us this property of matter is only affected by an external force. So, a

body will stay in a state of rest or will continue to keep moving if already doing so unless an external force is applied. If you imagine there being a toy car sat on your table, this remains stationary until you apply a force on it, for example by pushing it. As soon as you do push it, other forces start to act against it, the most notable being friction. It would travel a lot further if pushed in space because of there being fewer and weaker forces subsequently acting against it.

Second law

The alteration of motion is ever proportional to the motive force impressed; and is made in the direction of the right line in which that force is impressed.

This second law is best summarised by the equation $F = ma$, Force equals mass times acceleration. It's clear from this equation that the more force you apply, the greater the subsequent acceleration there will be. Imagine kicking a football with all your might compared with kicking it gently. It also tells us that the greater the mass something has, the greater the force that needs to be applied to get it to move. Imagine the difference between pushing a toy car and pushing a real one.

This brings us to the very important difference between mass and weight. Mass is measured in kilograms (kg) and is about how much stuff of something there is. It is the same wherever you are in the universe.

However, the weight of something is dependent on gravity and is measured in Newtons (N), which is the unit of force. So, based on our equation above, weight of an object with mass, m, corresponds to the equation W = ma. Here 'a' corresponds to gravitational acceleration, g, and the equation in this instance becomes W = mg. On Earth, g has the value of 9.8 m/s^2. On the moon it has the value 1.63 m/s^2. So, a person with a mass of 70kg would weigh 686 N on the Earth but on the moon would weigh about a sixth of this. So while the mass of something cannot change, the weight very much can.

Third law

To every Action there is always opposed an equal Reaction: or the mutual actions of two bodies upon each other are always equal, and directed to contrary parts.

It's no great revelation when I say the Earth's gravity is acting upon you all the time. It's what keeps our feet on the ground after all. But Newton's third law tells us that an equal (and opposing) force must also be happening. With this law, more than one object has to be involved. In the case of someone standing still, the ground is what is applying this. Another good example is a gun being fired. The bullet shoots out at great speed and, in return, the gun recoils. To prevent the gun from physically moving, an equal force has to be applied by the person firing the gun.

> *Blind commitment to a theory is not an intellectual virtue: it is an intellectual crime.*
>
> **Imre Lakatos, philosopher of science**

PERMISSIONS

Hans Van de Bovenkamp, executor for the Siv Cedering Estate, for allowing the reproduction of the poem 'Letter from Caroline Herschel (1750–1848)'.

Nielsen BookScan for providing sales information on popular science book sales from 2000–2009.

Theodor Benfey for allowing his periodic snail to be redrawn and reproduced.

'Preparing scientific papers' reprinted by permission from Macmillan Publishers Ltd: Nature 268, 1977.

Professor Paul May, for using material from his silly molecules website: www.chm.bris.ac.uk/sillymolecules/sillymols.htm.

'A Glossary for Research Reports' reprinted with permission of ASM International. All rights reserved. www.asminternational.org.

Professor Mike Benton for allowing his spindle diagram in *Vertebrate Palaeontology* to be redrawn and reproduced.

'Heaven is Hotter than Hell' reprinted by permission of The Optical Society.

The Exploratorium (www.exploratorium.edu) in San Francisco for allowing their forthcoming eclipse location map to be redrawn and reproduced.

Kate McAlpine (www.katemcalpine.com) for allowing her LHC lyrics to be reproduced.

'The Tale of Schrödinger's Cat' by Marilyn T. Kocher reprinted with permission from *Physics Today*, May 1978, American Institute of Physics.

ACKNOWLEDGEMENTS

'If we meet someone who owes us thanks, we think of it straightaway. But how often do we meet someone to whom we owe thanks without even thinking of it?'

Johann Wolfgang Von Goethe

This entire book is indebted to others, not least the many people mentioned throughout its pages and in the bibliography. Without them it couldn't exist, but there are many others to acknowledge too.

I worked for Icon Books, the publisher of *The Science Magpie*, for nearly fifteen years. During that time I was fortunate enough to be involved with a variety of fascinating, thought-provoking popular science titles. These – and of course their authors – have been a huge influence in writing this book and it has been a real pleasure to return to many of them in putting it together. Another significant inspiration has been the works of Professor R. L. Weber, particularly *A Random Walk in Science*. These wonderful physics anthologies are infused with an affectionate and playful love for their subject and have supplied a few of the items that have been included in this book. I strongly recommend you try to get hold of second-hand copies.

Given elsewhere is a list of permissions but regarding these, I'd like to thank (in some cases for the second time) Theodore Benfey, Hans Van de Bovenkamp,

Mike Keith, Paul May, Mike Benton, Kate McAlpine, Megan Bury, Hannah Bembia and Sue Sellers.

While the book was still being put together, Icon Books sent an email to staff interested in popular science at the bookseller Waterstones, asking for suggestions regarding material to include. Ideas came thick and fast and I was made aware of some wonderful new things, such as Edgar Allen Poe pre-empting the solution to Olber's Paradox. They also pointed me in new directions regarding some already drafted topics. A particularly big thank you to Emma Walton, Ashley Dolling, James Donaldson, Christian Delroy, Martin Keating, Michelle Wood and Melanie Jones.

Brian Clegg and John Waller read early versions of the text and offered corrections and advice, which were very gratefully received. I thank also my tried and trusted friends Rob Davies, Simon Politzer, Paul Sykes, Chris Tarling and Tamsin Tarling for their valuable input. Any remaining errors are down to me and me alone.

The team at Icon Books, along with their external colleagues, have been a joy to work with and have played a crucial role in the development of *The Science Magpie*. For a long time, the book had the working title *Science Miscellany*. In an early draft of the blurb I included the phrase 'This science miscellany presents a magpie's collection of ...' Andrew Furlow, Sales and Marketing Director, had the wisdom to see the potential in this and suggest *The Science Magpie* as the book's proper title – once that happened, its seemingly haphazard content all of a sudden made a lot more sense and

I am very grateful for his recognising this. Mark Ecob (www.mecob.org) came up with a quite brilliant cover design and Nick Halliday (www.hallidaybooks.com) worked wonders on the illustrations. Marie Doherty is a magical typesetter and the design she gave the book's insides felt immediately right. Sara Bryant is an intelligent and thorough (not to mention speedy!) proofreader. Duncan Heath, Editorial Director, has been extremely patient and understanding, as well as providing gentle and much appreciated humour throughout the writing of this book. Peter Pugh and Philip Cotterell's support for the book has never been short of 100%. A more positive and enthusiastic chairman and managing director of a publishing company an author couldn't wish for. Most of all, Harry Scoble, Editor, has, quite frankly, turned a sow's ear into … something better at least. No one should envy the task he was given and I only hope he's able to sleep more easily now – thank you!

Last, but most certainly not least, are Kate and Alice for their unfailing support. Without them, I would never have been able to write this book. The reason why is something that cannot be put into words but I hope they're aware of it.

BIBLIOGRAPHY

General resources consulted

Oxford Dictionary of National Biography, Oxford University Press, 2004

Oxford English Dictionary, Oxford University Press, accessed online 2012

Open University course materials S104 – *Exploring Science*
Open University course materials S205 – *The Molecular World*
Open University course materials MST121 – *Using Mathematics*
Open University course materials MS221 – *Exploring Mathematics*

The Times online archive
The *New York Times* online archive
The Examiner online archive
The *Illustrated London News* online archive
The *Morning Chronicle* archive
Morning Post and Daily Advertiser online archive
Science
Nature

wikipedia.com
wolframalpha.com
nobelprize.org

Sources regarding specific entries
Websites
http://www.jupiterscientific.org/sciinfo/jokes/miscellaneousjokes. html (How do they do IT?)
data.worldbank.org (Value judgements)
www.youtube.com/watch?v=1vBvp3MenVE (As easy as Al, Be, Cs)

http://www.british-history.ac.uk/report.aspx?compid=41125 (The square of scientific delights)

http://www.telegraph.co.uk/news/newstopics/howaboutthat/8971853/Skeleton-of-Charles-Byrne-the-Irish-Giant-should-be-buried-at-sea.html (The square of scientific delights)

darwin-online.org.uk/content/frameset?viewtype=text&itemID=CUL-DAR210.8.2&pageseq=1 (To marry or not to marry)

www.bbc.co.uk/news/uk-england-derbyshire-17315323 (Taking acid)

web.archive.org/web/19961031232918/http://media.circus.com/~no_dhmo/ (Ban dihydrogen monoxide!)

www.exploratorium.edu/eclipse/future.html (Solar eclipses galore)

http://privatewww.essex.ac.uk/~alan/family/N-Money.html ('A remarkable book, sure to make a mighty stir')

www.iucn.org/redlist/ (Threatened species)

http://www.bbc.co.uk/nature/extinction_events (The Big Five)

http://www.nature.com/nature/journal/v471/n7336/fig_tab/nature09678_T1.html (The Big Five)

impact.arc.nasa.gov/torino.cfm#how (The Torino impact hazard scale)

www.sothebys.com/en/auctions/2010/magnificent-books-manuscripts-and-drawings-from-the-collection-of-frederick-2nd-lord-hesketh-l10413/overview.html (The $10 million book)

www.economist.com/blogs/dailychart/2010/12/books (The $10 million book)

hub.salford.ac.uk/theoriesandmethodsliteraturescienceandmedicine/2011/01/06/sons-of-genius-by-sir-humphry-davy/ (The poets' scientist)

http://www.nature.com/scitable/topicpage/eukaryotic-genome-complexity-437 (When comes to what's in your genes, size isn't everything)

www.leeds.ac.uk/news/article/245/new_molecular_clock_aids_dating_of_human_migration_history (Would you Adam and Eve it?)

www.katemcalpine.com/scirap.html (The Large Hadron Collider rap)

www.modernlibrary.com/top-100/100-best-nonfiction/ (Modern Library's Top 100 non-fiction books)

www.physics.usyd.edu.au/~cross/Mariotte's%20Cradle.pdf (A law by any other name)

wordplay.blogs.nytimes.com/2010/12/06/numberplay-newtons-cradle/ (A law by any other name)

www.csgnetwork.com/h2oboilcalc.html (Just not my cup of tea)

ebooks.adelaide.edu.au/d/darwin/charles/beagle/chapter15.html (Just not my cup of tea)

www.chm.bris.ac.uk/sillymolecules/sillymols.htm (Real molecules with silly names)

www.stsci.edu/institute/smo/visitor-programs/caroline-herschel/poem (Britain's first professional female scientist)

www.iau.org/public/pluto/ (The definition of a planet)

www.nasa.gov/centers/ames/missions/archive/pioneer.html (The interstellar pioneer)

grin.hq.nasa.gov/ABSTRACTS/GPN-2000-001623.html (The interstellar pioneer)

coolcosmos.ipac.caltech.edu/cosmic_classroom/classroom_activities/herschel_bio.html (Another hue unto the rainbow)

news.bbc.co.uk/1/hi/health/8339877.stm (Centenary icons of science and technology)

royalsociety.org/awards/science-books/ (Royal Society prize for science books)

www.scientificamerican.com/article.cfm?id=kuhn (The book that caused a paradigm shift)

www.guardian.co.uk/news/datablog/2011/mar/15/radiation-exposure-levels-guide#data (The dose makes the poison)

www.hpa.org.uk/Topics/Radiation/UnderstandingRadiation/UnderstandingRadiationTopics/DoseComparisonsFor IonisingRadiation (The dose makes the poison)

www.world-nuclear.org/education/ral.htm (The dose makes the poison)

www.world-nuclear.org/info/inf05.html (The dose makes the
poison)

www.bbc.co.uk/news/health-12722435 (The dose makes the
poison)

www.atomicarchive.com/History/firstpile/firstpile_01.shtml#2
(Going nuclear in a sports hall)

www.cddc.vt.edu/host/atomic/trinity/trinity1.html (An unholy
trinity)

plato.stanford.edu/entries/qm-copenhagen/ (Would the real
Schrödinger's cat please stand up)

www.lablit.com/the_list (Truth to their fictions)

www.agecon.purdue.edu/crd/Localgov/Second%20Level%20
pages/indiana_pi_bill.htm (A slice of Indiana π)

www.agecon.purdue.edu/crd/Localgov/Second%20Level%20
pages/Indiana_Pi_Story.htm (A slice of Indiana π)

www.newtonproject.sussex.ac.uk/prism.php?id=135&name=53
(Meet Tom Telescope and friends)

www.newtonproject.sussex.ac.uk/view/texts/normalized/
ALCH00069 (The sins of Newton)

http://law2.umkc.edu/faculty/projects/ftrials/scopes/statcase.htm
(Monkey Trial)

http://www.tn.gov/tsla/exhibits/scopes/index.htm (Monkey Trial)

science.ksc.nasa.gov/shuttle/technology/sts-newsref/srb.html
(Five, four, three, two one... we have lift off)

http://www.ncstatecollege.edu/Webpub/kekegren/enr280f00/
Lesson11a.htm (Five, four, three, two one... we have lift off)

www.nlm.nih.gov/hmd/greek/greek_oath.html (First do no harm)

gnomeexperiment.com (The size and gravity of the situation)

kernthegnome.tumblr.com (The size and gravity of the situation)

www.columbia.edu/~gas1/project/visions/case1/sci.2.html
(An unacknowledged debt)

www.food.gov.uk/policy-advice/additivesbranch/enumberlist
(E's are... additives)

www-users.cs.york.ac.uk/susan/joke/exoendo.htm (Is hell
exothermic or endothermic?)

alchemipedia.blogspot.co.uk/2009/12/100-greatest-britons-
     bbc-poll-2002.html (Greatest scientists by nation)
www.filibustercartoons.com/greatest%20Canadians.htm (Greatest
     scientists by nation)
www.ceskatelevize.cz/specialy/nejvetsicech/oprojektu_top100
     (Greatest scientists by nation)
knarf.english.upenn.edu/Darwin/templetp.html ('Successive
     generations bloom')

Books/articles

Oxford Dictionary of Science, Oxford University Press, Oxford, 2010
'University of Cambridge Local Examinations Syndicate question
     papers, timetable, grace, regulations and notices, 1858'
'University of Cambridge Local Examinations Syndicate First
     Annual Report, 1859'

Anonymous, 'Heaven is Hotter than Hell', *Applied Optics*, 11,
     1972
Armstrong, Isobel, Bristow, Joseph and Sharrock, Cath, eds.
     Nineteenth-Century Women Poets, Oxford University Press,
     Oxford, 1996
Benfey, Theodor, 'The Biography of a Periodic Snail', *Bulletin of
     the History of Chemistry*, Vol. 34, No. 2
Bizony, Piers, *Atom*, Icon Books, London, 2007
Blakemore, Colin and Jennett, Sheila, eds. *The Oxford Companion
     to the Body, Oxford University Press*, Oxford, 2001
Brock, Claire, *The Comet Sweeper*, Icon Books, London, 2007
Bruce, Susan, *Three Early Modern Utopias*, Oxford University Press,
     Oxford, 1999
Burrows, Andrew, Holman, John, Parsons, Andrew, Piling, Gwen
     and Price, Gareth, *Chemistry³*, Oxford University Press,
     Oxford, 2009
Bynum, W. F., and Porter, Roy, *Oxford Dictionary of Scientific
     Quotations*, Oxford University Press, Oxford, 2005

Campbell, Lewis and Garnett, William, *The Life of James Clerk Maxwell*, Macmillan and Co., 1882

Carey, Nessa, *The Epigenetics Revolution*, Icon Books, London, 2011

Chalmers, A. F., *What is this thing called Science?*, Open University Press, Milton Keynes, 1999

Cheater, Christine, 'Collectors of Nature's Curiosities' in *Frankenstein's Science*, Ashgate, Aldershot, 2008

Clarke, S. H., *Akenside, Macpherson & Young*, Carcanet Press, Manchester, 1994

Clegg, Brian, *Inflight Science*, Icon Books, 2011

Clerk Maxwell, James, *Theory of Heat*, Longmans, Green, and Co., London, 1902

Connolly, Sean, *The Book of Potentially Catastrophic Science*, Icon Books, London, 2011

Cruciani, Fulvio, Trombetta, Beniamino, Massaia, Andrea, Destro-Bisol, Giovanni, Sellitto, Daniele, Scozzari, Rosaria, 'A Revised Root for the Human Y Chromosomal Phylogenetic Tree: The Origin of Patrilineal Diversity in Africa', *American Journal of Human Genetics*, Vol. 88 No. 6, 2011

Darwin, Charles, *A Naturalist's Voyage Round the World*, John Murray, London, 1913

Darwin, Charles, *On the Origin of Species*, Penguin Books, London, 2009

Davies, B, 'Edme Mariotte 1610–1684', *Physics Education* Vol. 9 No. 275, 1974

Einstein, Albert, *Autobiographical Notes*, Open Court Publishing, Peru, 1979

Ellegård, Alvar, *Darwin and the General Reader*, 1859-1872, Göteborg, Stockholm, 1958

Elsdon-Baker, Fern, *The Selfish Genius*, Icon Books, London, 2009

Engels, David, 'The Length of Eratosthenes' Stade', *The American Journal of Philology*, Vol. 106, No. 3 (Autumn, 1985)

Fara, Patricia, *Scientists Anonymous*, Wizard Books, London, 2005

Feyerabend, Paul, 'Patterns of Discovery', *The Philosophical Review*, Vol. 69 No. 2, 1960

Feynman, Richard, *Six Easy Pieces*, Penguin Books, London, 1995

Flood, W. E., *Scientific Words*, The Scientific Book Guild, 1961

Foot, Paul, *Red Shelley*, Bookmarks, London, 1984

Graham, Jr, C. D., 'A Glossary for Research Reports', *Metal Progress*, Vol. 71 No. 5, 1957

Gregory, Andrew, *Eureka*, Icon Books, 2001

Gullberg, Jan, *Mathematics: From the Birth of Numbers*, W. W. Norton & Company, New York, 1997

Haile, N. S., 'Preparing Scientific Papers', *Nature*, Vol. 268, 1977

Hannam, James, *God's Philosophers*, Icon Books, London, 2009

Hanson, Norwood Russell, *Patterns of Discovery*, Cambridge University Press, Cambridge, 1965

Hart-Davis, Adam, ed. *Science*, Dorling Kindersley, London, 2009

Haynes, Clare, 'A Natural Exhibitioner: Sir Ashton Lever and his Holophusikon' *British Journal for Eighteenth-Century Studies* 24, 2001

Hazen, Robert M. and Trefil, James, *Science Matters*, Anchor Books, New York, 2009

Henry, John, *Knowledge is Power*, Icon Books, London, 2002

Henry, John, *Moving Heaven and Earth*, Icon Books, London, 2001

Hogg, Thomas Jefferson, 'Shelley at Oxford', *New Monthly Magazine*, 1832

Huxley, Leonard, *Life and Letters of Thomas Henry Huxley Vol 1*

James, Arthur M. and Lord, Mary P., *Macmillan's Chemical and Physical Data*, The Macmillan Press, London, 1992

Jargodzki, Christopher and Potter, Franklin, *Mad About Physics*, John Wiley & Sons, Inc., 2000

Jargodzki, Christopher and Potter, Franklin, *Mad About Modern Physics*, John Wiley & Sons, Inc., 2005

Jones, Bence, *The Life and Letters of Faraday*, Longmans, Green, and Co., London, 1870

Kipperman, Mark, 'Coleridge, Shelley, Davy, and Science's Millennium', *Criticism*, 40:3, 1998: Summer

Knight, David, 'Humphry Davy the Poet', *Interdisciplinary Science Reviews*, Vol. 30, No. 4, 2005

Knowles, Elizabeth, ed. *The Oxford Dictionary of Quotations*, Oxford University Press, Oxford, 1999

Kuhn, Thomas S., *The Structure of Scientific Revolutions*, University of Chicago Press, Chicago, 2012

Kumar, Manjit, *Quantum*, Icon Books, London, 2008

Ladd, Everett Carll, Jr. and Lipset, Seymour Martin, 'Politics of Academic Natural Scientists and Engineers', *Science*, Vol. 176, No. 4039, 1972

Lamont-Brown, Raymond, *Humphry Davy*, Sutton Publishing, Stroud, 2004

Mackay, Alan L, *A Dictionary of Scientific Quotations*, IOP Publishing, Bristol, 1991

Maor, Eli, *e: the Story of a Number*, Princeton University Press, Princeton, 1994

Masood, Ehsan, *Science & Islam*, Icon Books, London, 2009

McEvoy, J.P. and Zarate, Oscar, *Introducing Quantum Theory*, Icon Books, London, 1997

Mpemba, E. B. and Osborne, D. G., 'Cool?', *Physics Education*, Vol 4, 1969

Nelson, David, ed. *The Penguin Dictionary of Mathematics*, Penguin Books, 2003

Newton, Isaac, *The Mathematical Principles of Natural Philosophy* Volume 1, Benjamin Motte, London, 1729

OECD, *Science, Technology and Industry Outlook*, 2010, p. 43

Perez, Jorge Mira and Vina, Jose, 'Physics, Bible Used to Re-examine if Heaven Is Hotter than Hell', *Physics Today*, vol. 51(7), 1998

Pickover, Clifford A., *The Math Book*, Sterling Publishing, New York, 2009

Pickover, Clifford A., *A Passion for Mathematics*, John Wiley & Sons, Hoboken, 2005

Poe, Edgar Allen, *Eureka: A Prose Poem*, Geo. P. Putnam, New York, 1848

Porter, Roy, ed. *The Cambridge Illustrated History of Medicine*, Cambridge University Press, Cambridge, 1996

Pray, L., Eukaryotic Genome Complexity, *Nature Education*, 1(1)

Ragep, F. Jamil, 'Copernicus and his Islamic Predecessors', *History of Science*, xlv, 2007

Rhys Morus, Iwan, *Michael Faraday and the Electrical Century*, Icon Books, London, 2004

Riordan, Maurice and Turney, Jon, eds. *A Quark for Mister Mark*, Faber, London, 2000

Schrödinger, Erwin, 'Die gegenwärtige Situation in der Quantenmechanik', *Naturwissenschaften* 23, translated by Trimmer, John D., in *Proceedings of the American Philosophical Society*, 124

Secord, Jim, 'Newton in the Nursery', *History of Science*, xxiii, 1985

Snow, C. P., *The Two Cultures*, University of Cambridge, Cambridge, 1993

Tange, Andrea Kaston, 'Constance Naden and the Erotics of Evolution: Mating the Woman of Letters with the Man of Science', *Nineteenth-Century Literature*, Vol. 61, No. 2 (September 2006)

Tennent, R. M., ed. *Science Data Book*, Oliver & Boyd, 1971

Timbs, John, *The Year Book of Facts in Science and Arts 1855*, David Bogue, London, 1855

Toyabe, S., Sagawa, T., Ueda, M., Muneyuki, E. & Sano, M., *Nature Physics*, Vol. 6, 2010

Waller, John, *The Discovery of the Germ*, Icon Books, London, 2002

Waller, John, *Fabulous Science*, Oxford University Press, Oxford, 2002

Weber, R. L., comp. *A Random Walk in Science*, IOP Publishing, London, 1973

Weber, R. L., comp. *Droll Science*, Humana Press, Clifton, 1987

Weber, R. L., comp. *More Random Walks in Science*, IOP Publishing, London, 1982

Weber, R. L., comp. *Science with a Smile*, IOP Publishing, London, 1992

White, William, *The Illustrated Hand Book of the Royal Panopticon of Science and Art*, John Hotson, London, 1854

Wyatt, John, *Wordsworth and the Geologists*, Cambridge University Press, Cambridge, 1995

THE SCIENCE MAGPIE